Giuliano Armano, Marco de Gemmis, Giovanni Semeraro, and Eloisa Vargiu (Eds.)

Intelligent Information Access

Studies in Computational Intelligence, Volume 301

Editor-in-Chief
Prof. Janusz Kacprzyk
Systems Research Institute
Polish Academy of Sciences
ul. Newelska 6
01-447 Warsaw
Poland
E-mail: kacprzyk@ibspan.waw.pl

Further volumes of this series can be found on our homepage: springer.com

Vol. 279. Manolis Wallace, Ioannis E. Anagnostopoulos, Phivos Mylonas, and Maria Bielikova (Eds.)
Semantics in Adaptive and Personalized Services, 2010
ISBN 978-3-642-11683-4

Vol. 280. Chang Wen Chen, Zhu Li, and Shiguo Lian (Eds.)
Intelligent Multimedia Communication: Techniques and Applications, 2010
ISBN 978-3-642-11685-8

Vol. 281. Robert Babuska and Frans C.A. Groen (Eds.)
Interactive Collaborative Information Systems, 2010
ISBN 978-3-642-11687-2

Vol. 282. Husrev Taha Sencar, Sergio Velastin, Nikolaos Nikolaidis, and Shiguo Lian (Eds.)
Intelligent Multimedia Analysis for Security Applications, 2010
ISBN 978-3-642-11754-1

Vol. 283. Ngoc Thanh Nguyen, Radoslaw Katarzyniak, and Shyi-Ming Chen (Eds.)
Advances in Intelligent Information and Database Systems, 2010
ISBN 978-3-642-12089-3

Vol. 284. Juan R. González, David Alejandro Pelta, Carlos Cruz, Germán Terrazas, and Natalio Krasnogor (Eds.)
Nature Inspired Cooperative Strategies for Optimization (NICSO 2010), 2010
ISBN 978-3-642-12537-9

Vol. 285. Roberto Cipolla, Sebastiano Battiato, and Giovanni Maria Farinella (Eds.)
Computer Vision, 2010
ISBN 978-3-642-12847-9

Vol. 286. Zeev Volkovich, Alexander Bolshoy, Valery Kirzhner, and Zeev Barzily
Genome Clustering, 2010
ISBN 978-3-642-12951-3

Vol. 287. Dan Schonfeld, Caifeng Shan, Dacheng Tao, and Liang Wang (Eds.)
Video Search and Mining, 2010
ISBN 978-3-642-12899-8

Vol. 288. I-Hsien Ting, Hui-Ju Wu, Tien-Hwa Ho (Eds.)
Mining and Analyzing Social Networks, 2010
ISBN 978-3-642-13421-0

Vol. 289. Anne Håkansson, Ronald Hartung, and Ngoc Thanh Nguyen (Eds.)
Agent and Multi-agent Technology for Internet and Enterprise Systems, 2010
ISBN 978-3-642-13525-5

Vol. 290. Weiliang Xu and John Bronlund
Mastication Robots, 2010
ISBN 978-3-540-93902-3

Vol. 291. Shimon Whiteson
Adaptive Representations for Reinforcement Learning, 2010
ISBN 978-3-642-13931-4

Vol. 292. Fabrice Guillet, Gilbert Ritschard, Henri Briand, Djamel A. Zighed (Eds.)
Advances in Knowledge Discovery and Management, 2010
ISBN 978-3-642-00579-4

Vol. 293. Anthony Brabazon, Michael O'Neill, and Dietmar Maringer (Eds.)
Natural Computing in Computational Finance, 2010
ISBN 978-3-642-13949-9

Vol. 294. Manuel F.M. Barros, Jorge M.C. Guilherme, and Nuno C.G. Horta
Analog Circuits and Systems Optimization based on Evolutionary Computation Techniques, 2010
ISBN 978-3-642-12345-0

Vol. 295. Roger Lee (Ed.)
Software Engineering, Artificial Intelligence, Networking and Parallel/Distributed Computing, 2010
ISBN 978-3-642-13264-3

Vol. 296. Roger Lee (Ed.)
Software Engineering Research, Management and Applications, 2010
ISBN 978-3-642-13272-8

Vol. 297. Tania Tronco (Ed.)
New Network Architectures, 2010
ISBN 978-3-642-13246-9

Vol. 298. Adam Wierzbicki
Trust and Fairness in Open, Distributed Systems, 2010
ISBN 978-3-642-13450-0

Vol. 299. Vassil Sgurev, Mincho Hadjiski, and Janusz Kacprzyk (Eds.)
Intelligent Systems: From Theory to Practice, 2010
ISBN 978-3-642-13427-2

Vol. 300. Baoding Liu (Ed.)
Uncertainty Theory, 2010
ISBN 978-3-642-13958-1

Vol. 301. Giuliano Armano, Marco de Gemmis, Giovanni Semeraro, and Eloisa Vargiu (Eds.)
Intelligent Information Access, 2010
ISBN 978-3-642-13999-4

Giuliano Armano, Marco de Gemmis,
Giovanni Semeraro, and Eloisa Vargiu (Eds.)

Intelligent Information Access

Springer

Giuliano Armano
Department of Electrical and
Electronic Engineering
University of Cagliari
Piazza d'Armi - I09123 Cagliari
Italy
E-mail: armano@diee.unica.it

Giovanni Semeraro
Department of Informatics
University of Bari "Aldo Moro"
Via E. Orabona, 4
70126 - Bari
Italy
E-mail: semeraro@di.uniba.it

Marco de Gemmis
Department of Informatics
University of Bari "Aldo Moro"
Via E. Orabona, 4
70126 - Bari
Italy
E-mail : degemmis@di.uniba.it

Eloisa Vargiu
Department of Electrical and
Electronic Engineering
University of Cagliari
Piazza d'Armi - I09123 Cagliari
Italy
E-mail: vargiu@diee.unica.it

ISBN 978-3-642-13999-4 e-ISBN 978-3-642-14000-6

DOI 10.1007/978-3-642-14000-6

Studies in Computational Intelligence ISSN 1860-949X

Library of Congress Control Number: 2010929476

Typeset & Cover Design: Scientific Publishing Services Pvt. Ltd., Chennai, India.

Printed on acid-free paper

9 8 7 6 5 4 3 2 1

springer.com

Preface

Intelligent Information Access techniques attempt to overcome the limitations of current search devices by providing personalized information items and product/service recommendations. They normally utilize direct or indirect user input and facilitate the information search and decision processes, according to user needs, preferences and usage patterns. Recent developments at the intersection of Information Retrieval, Information Filtering, Machine Learning, User Modelling, Natural Language Processing and Human-Computer Interaction offer novel solutions that empower users to go beyond single-session lookup tasks and that aim at serving the more complex requirement: "Tell me what I don't know that I need to know". Information filtering systems, specifically recommender systems, have been revolutionizing the way information seekers find what they want, because they effectively prune large information spaces and help users in selecting items that best meet their needs and preferences. Recommender systems rely strongly on the use of various machine learning tools and algorithms for learning how to rank, or predict user evaluation, of items. Information Retrieval systems, on the other hand, also attempt to address similar filtering and ranking problems for pieces of information such as links, pages, and documents. But they generally focus on the development of global retrieval techniques, often neglecting individual user needs and preferences.

The book aims to investigate current developments and new insights into methods, techniques and technologies for intelligent information access from a multidisciplinary perspective. It comprises six chapters authored by participants in the research event *Intelligent Information Access*, held in Cagliari (Italy) in December 2008.

In Chapter 1, *Enhancing Conversational Access to Information through a Socially Intelligent Agent*, Berardina De Carolis, Irene Mazzotta and Nicole Novielli emphasize the role of Embodied Conversational Agents (ECAs) as a natural interaction metaphor for personalized and context-adapted access to information. They propose a scalable architecture for the development of ECAs able to exhibit an emotional state and/or social signs.

The automatic detection of emotions in text is the problem investigated in Chapter 2, *Annotating and Identifying Emotions in Text*, by Carlo Strapparava and Rada Mihalcea. The authors describe the "Affective Text" task, presented at SEMEVAL- 2007. The task focused on classifying emotions in news headlines, and was intended to explore the connection between emotions and lexical semantics. After illustrating the data set, the rationale of the task and a brief description of the participating systems, several experiments on the automatic annotation of emotions in text are presented. The practical applications of the task are very important. Consider for example opinion mining and market analysis, affective computing, natural language interfaces for e-learning environments or educational games.

Personalization of the ranking computed by search engines and recommender systems is the main topic of Chapter 3, *Improving Ranking by Respecting the Multidimensionality and Uncertainty of User Preferences*, by Bettina Berendt and Veit Koppen. The research question addressed by the authors is whether system ranking is the "right ranking" for the user, based on the context in which she/he operates. A general conceptualization of the ranking-evaluation task is proposed: the comparison between the ranking generated by a computational system, and the " user's ideal ranking". Eight challenges to this simple model are discussed, leading to the conclusion that approaches for dealing with multidimensional, and often only partial, preference orders are required and that randomness could be a beneficial feature of system rankings.

In Chapter 4, Hotho reviews the state of the art in the new research area of *data mining on folksonomies*. The first part describes the basics of folksonomies, summarizing del.icio.us, the most popular social bookmarking system, and illustrates in detail BibSonomy, a very successful online service for social bookmarking and publication sharing. Starting from these systems, the author discusses in greater depth the main issues regarding folksonomies, proposing a formal model and presenting their most important network properties. In the second part, the author illustrates three applications: spam detection, ranking and recommendation. Regarding spam detection, the author develops techniques, based on binary classifiers, which prevent spammers from publishing in social bookmarking systems. As far as ranking is concerned, a new algorithm is proposed, namely FolkRank, which takes into account the folksonomy structure for ranking users, tags and resources. For recommendation, the author evaluates a tag recommender based on Collaborative Filtering, a graph based recommender using FolkRank and several simple approaches based on tag counts. In the third part, a possible link between folksonomies and ontologies is suggested, paving the way to some very promising strategies for detecting organizational principles hidden within folksonomies.

Amati, Amodeo, Bianchi, Gaibisso and Gambosi propose, in Chapter 5, *A Uniform Theoretic Approach to Opinion and Information Retrieval*, an application of the Divergence From Randomness (DFR) model to the opinion finding task, the task of retrieving opinionated blog posts, relevant for a given topic, from a large collection. The opinion finding task can be seen as a search in which, after the standard retrieval of ranked documents, documents are re-ranked according to the presence of opinions within the selected documents. This task can be handled by a supervised or unsupervised method. The authors propose a method for creating a lexicon of opinionated terms for re-ranking the documents, using a supervised algorithm. The first part introduces the statistical basis underpinning the proposed approach and its adoption in opinion retrieval. In particular, two information-theoretic functions are defined, opinion entropy and average opinion entropy. The authors also formally describe their lightweight opinion retrieval algorithm. Lastly, the authors discuss the effectiveness of their approach for creating a dictionary of polarity-bearing terms. They also describe some preliminary experiments and propose alternative ways to approach the polarity detection problem.

In Chapter 6, *A Suite of Semantic Web Tools Supporting Development of Multilingual Ontologies*, Pazienza, Stellato and Turbati propose a suite of software libraries, tools and ontologies to support multilingual development of Semantic Web ontologies. The three tools illustrated in this Chapter are Semantic Turkey, The Linguistic Watermark, and Ontoling. Semantic Turkey is aimed at providing innovative solutions for web browsing and for gathering and organizing the information observed when surfing the net. The novel aspect of Semantic Turkey is its ability to provide a clear separation between acquired data and web links. The Linguistic Watermark is an ontological and software framework for describing and managing heterogeneous linguistic resources and for using their contents for ontological-driven document enrichment. Ontoling is a generic architecture for extending ontology development tools with functionalities for enriching ontological knowledge with linguistic content. The tools presented implicitly embed a new way of rethinking the development of ontologies in terms of making their content reusable and comprehensible. Furthermore, they represent living proof of software engineering principles associated with software reuse, documentation, modularity, interaction analysis, applied to the domain of Knowledge Management Software.

We would like to thank all the authors for their excellent contributions and the reviewers for their careful revision and suggestions for improving them. We are grateful to the Springer-Verlag Team for their assistance during the preparation of the manuscripts.

This book is dedicated to the memory of Fiorella de Rosis in recognition of her contribution to user modeling. She was a pioneer in the field of affective computing, a leader in research on modeling emotions and constructing embodied animated agents. She produced key contributions in intelligent user interfaces, in particular on user-adapted generation of natural language and multimedia messages, uncertainty in user models, and presentation of medical explanations and clinical guidelines. During her teaching and research activities she mentored many students who have become established researchers. These research and teaching activities didn't prevent her from being an active member of the ACM, of the International Society for Research on Emotions, of the European Network of Excellence on Emotions (HUMAINE), of the editorial boards of UMUAI and co-chair of many international conferences. All the people acquainted with Fiorella have appreciated her scientific and human value and are grateful for her friendship.

February 2010

Giuliano Armano
Marco de Gemmis
Giovanni Semeraro
Eloisa Vargiu

Contents

List of Contributors

Giambattista Amati
Fondazione Ugo Bordoni,
Rome, Italy
gba@fub.it

Giuseppe Amodeo
Dept. of Computer Science,
University of L'Aquila, Italy
gamodeo@fub.it

Bettina Berendt
K.U. Leuven,
Dept. of Computer Science,
Leuven, Belgium
http://www.cs.kuleuven.be/ berendt

Marco Bianchi
Istituto di Analisi dei Sistemi
ed Informatica
"Antonio Ruberti" - CNR,
Rome, Italy
surname@iasi.cnr.it

Berardina De Carolis
Dipartimento di Informatica,
Università degli Studi di Bari,
Via Orabona,
4 - 70125 - Bari, Italy
decarolis@di.uniba.it

Carlo Gaibisso
Istituto di Analisi dei
Sistemi ed Informatica
"Antonio Ruberti" - CNR,
Rome, Italy
surname@iasi.cnr.it

Giorgio Gambosi
Dept. of Mathematics,
University of Rome
"Tor Vergata", Italy,
gambosi@mat.uniroma2.it

Andreas Hotho
Knowledge & Data
Engineering Group,
University of Kassel,
34121 Kassel, Germany
hotho@cs.uni-kassel.de

Veit Köppen
Otto-von-Guericke-Universität
Magdeburg,
Dept. of Technical & Business
Information Systems,
Magdeburg, Germany
http://www.veit-koeppen.de/

Irene Mazzotta
Dipartimento di Informatica,
Università degli Studi di Bari,

Via Orabona,
4 - 70125 - Bari, Italy
mazzotta@di.uniba.it

Rada Mihalcea
University of North Texas
rada@cs.unt.edu

Nicole Novielli
Dipartimento di Informatica,
Università degli Studi di Bari,
Via Orabona,
4 - 70125 - Bari,
Italy
novielli@di.uniba.it

Maria Teresa Pazienza
ART Group,
Dept. of Computer Science,
Systems and Production
University of Rome,
Tor Vergata
Via del Politecnico 1,
00133 Rome, Italy
pazienza@info.uniroma2.it

Armando Stellato
ART Group,
Dept. of Computer Science,
Systems and Production
University of Rome,
Tor Vergata
Via del Politecnico 1,
00133 Rome, Italy
stellato@info.uniroma2.it

Carlo Strapparava
FBK-IRST
strappa@fbk.edu

Andrea Turbati
ART Group,
Dept. of Computer Science,
Systems and Production
University of Rome,
Tor Vergata
Via del Politecnico 1,
00133 Rome, Italy
turbati@info.uniroma2.it

Enhancing Conversational Access to Information through a Socially Intelligent Agent

Berardina De Carolis, Irene Mazzotta, and Nicole Novielli

Abstract. Intelligent access to information could benefit of an effective and natural interaction metaphor. In this perspective, Embodied Conversational Agents (ECAs) can be seen as a promising approach to give to the user the illusion of cooperating with a partner rather than just using a tool. Embedding the HCI technology with human preferences and behavior justifies the attempt of implementing emotional and social intelligence aimed at exceeding the single ability to help the user. In this paper we present an ECA's architecture and methods useful to interpret the user attitude during her dialog with an ECA and behaving 'believably' in its turn. In particular, we present an agent architecture that is general enough to be applied in several application domains and that employs several ECA's bodies according to the context requirements.

Keywords: Natural Language Interaction, Conversational Access to Information, Emotional Intelligence, Embodied Conversational Agents.

1 Introduction

Intelligent Information Access has the main goal of providing a personalized access to information by exploiting information retrieval techniques. To this aim, the user behavior is observed, either directly or indirectly, in order to build user models which allow personalization (Kobsa 1993, Berkovsky et al. 2009).

We believe that a system providing intelligent and personalized information access, though, surely benefits of an intelligent presentation of the information content by exploiting methods for developing effective, usable and natural interaction metaphors.

An intelligent information system should therefore be equipped with and intelligent interface, that is an interface able to:

- adapt to the user;
- handle natural language dialogs using the appropriate strategies;
- decide, autonomously, when to activate themselves and how to respond to the (presumed) user needs;

Berardina De Carolis · Irene Mazzotta · Nicole Novielli
Dipartimento di Informatica, Università degli Studi di Bari,
Via Orabona, 4 - 70125 - Bari, Italy
e-mail: {decarolis,mazzotta,novielli}@di.uniba.it

G. Armano et al. (Eds.): Intelligent Information Access, SCI 301, pp. 1–20.
springerlink.com

- consider user emotions and social attitude during the dialogs and behave 'believably' in its turn.

In this perspective, Embodied Conversational Agents (ECA) can be seen as a new metaphor of human-computer ‘intelligent’ interaction which promises to be effective (Cassell and Bickmore 2003) if the hypothesis that ‘characters contribute to more sociable and user-friendly interfaces’ is taken for granted (Lee and Nass 1999). A well designed ECA should give the users the illusion of cooperating with a human partner rather than just ‘using a tool’. The more the agent succeeds in this goal the more the users are expected to attach some anthropomorphic features to them and to show signs of affective (emotional, social) involvement in the interaction. Therefore, in developing a "computer conversationalist" that is embedded in a ECA and that is able to exhibit these capabilities it is important to conceive its architecture so as to:

1. start from the interpretation of the spoken or written utterance;
2. reason on the various information the user intends to convey (emotion, social attitude, performative, content, etc.) and then to trigger communicative goals according to the current belief representation of the state of the world;
3. achieve these goals through a set of communicative plans (“what to say”) that can then be rendered as a combination of voice and animations of the agent’s body (“how to say”).

To achieve believable natural language conversations, ECA systems are quite complex and require the combination of several assemblies (Huang et al., 2008). In recent years, the members of the international research community have been jointly working towards the definition of a standard framework for the generation of behavior of virtual agents. The SAIBA project (Vilhjálmsson et al., 2007) represents one of the major efforts towards the unification of ECA standards and aims at unifying the key interfaces in the multimodal behavior generation process. In particular, the SAIBA architecture is structured on three different layers, each of them implemented to serve to a different function of the agent behavior planning and realization. Moreover, as far as affective computing is concerned, the SEMAINE API provides an open source framework for building systems embedding emotional intelligence: the SEMAINE system is a full-scale system resulting from the integration of several existing and new components (Schröder, 2010).

The example provided by these projects demonstrates how scalability and openness represent two extremely important challenges in developing architectures for believable ECAs: as far as scalability and openness are ensured, standard architectures can be easily extended to meet the specific needs of the various interaction scenarios and application domains (Bevacqua et al., 2009). The architecture proposed in the present paper perfectly fits in the vein of this ongoing research.

In particular, this paper describes our experience in the design and implementation of a scalable architecture of a believable ECA that interacts with the user for providing advices in a domain where considering social and affective factors is crucial. To this aim, the architecture is designed so as to dynamically model and build different agent’s functionalities, according to the application domain needs.

Moreover, taking into account context factors, it allows: i) employing the most appropriate dialog strategy (for instance, information giving vs. persuasion dialogs), ii) simulating more or less rational or affective agent's behaviors (Carofiglio et al. 2009), iii) using different agent's bodies.

The paper is structured as follows: Section 2 describes the agent architecture; then in Section 3 we provide an overview of the Interpersonal Stances Modeling; Section 4 presents a Model of Emotion Activation. In Section 5 we describe the Dialog Modeling component with an example of implemented dialog (Section 6). Section 7 shows a possible extension of our agent architecture, by describing how a persuasion module can be integrated and used opportunely in the agent architecture. Conclusions and future work directions are presented in the last Section.

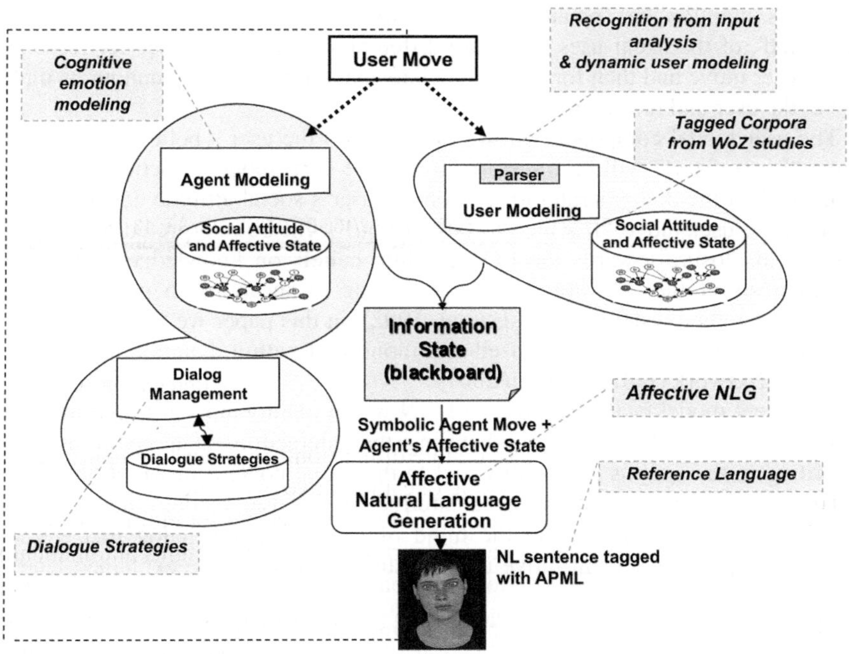

Fig. 1 An overview of the ECA architecture

2 The Architecture

The ability to exhibit an emotional state and/or social signs is a shallow form of the intelligence an agent can show. The recognition of the social attitude and of the emotional state of the interlocutor should be utilized to drive reasoning behind the dialog between the user and the ECA. This implies studying how these factors may affect the ECA architecture. In our opinion, when developing an ECA, the following issues should be addressed:

- the user move should be interpreted so as to detect, beside the linguistic content: i) which is the social attitude of the user and ii) which emotions arise during the dialog;
- how these factors influence the dialog course by changing the priority of communicative goals, dialog plan and surface realization of communicative acts.

Figure 1 illustrates the architecture we propose to handle these issues. This architecture has been conceived as composed by two main functional modules: the "mind" and the "body" of the agent.

2.1 The Mind

The user move is a rich information source that allows extracting knowledge about the user's intention, her social attitude, emotional state, and so on. In our approach, the "mind" of the agent uses two main different knowledge sources for reasoning on the user move and then formalizing beliefs that are useful for planning its dialog move: the user and the agent models.

The user model component allows to reason on the user's beliefs (i.e. the user move "I love fruit!" will be transformed into the correspondent belief that can be used to adapt the dialog strategy) and on the user's social attitude during the dialog (i.e. the user move "It's nice to talk to you!" will be interpreted as a sign of friendly disclosure towards the ECA). While beliefs on knowledge, preferences and interests of the user are inferred according to an approach previously employed in another system (de Rosis et al. 1992), in this paper we will explain how the user social attitude is recognized and monitored with a dynamic model based on Belief Network (DBN) (Jensen 2001).

The agent model is also based on a DBN which mainly aims at triggering emotions that arise in the agent mind during the interaction, in a given situation, according to the agent's personality and to the social context in which the dialog occurs. Starting from what has been inferred by the user model component and from the emotions triggered in the mind of the agent, the dialog management module computes the agent move using a strategy that will be explained later in the paper.

The information exchange among these modules is managed using a common blackboard called information state (Larsson and Traum 2000). It represents the memory of the agent and stores beliefs about the current state of the dialog, the dialog history, the current dialog move and the move scheduled for execution. This approach allows employing different methods and techniques giving to the architecture a degree of openness and scalability.

2.2 The Body

While the move computed by the "mind" module contains the meaning to express ("what to say"), the "body" has to convey these meaning according to its communicative capabilities ("how to say"). In order to decouple meanings from signals we use a mark-up language: APML (De Carolis et al. 2004). These meanings include the communicative functions that are typically used in human-human dialogs: for instance, syntactic, dialogic, meta-cognitive, performative, affective, deictic, adjectival and belief relation functions (Poggi et al. 2000).

The use of a reference language gives the possibility to employ different bodies and different platforms and devices without changing the mind of the agent. In fact, in order to express the same meaning using different signals according for instance to the context or to the capabilities of the body of the employed ECA, each ECA's body has a conditional meaning-signal table that allows to appropriately translate an APML tag into tags expressed in Signal Expression Markup Language (SEML). SEML tags define the expressions that can be performed on each channel of the Body as described in (De Carolis 2005).

Let's see now in more details how these modules work.

3 Interpersonal Stances Modeling

After several forms of 'anthropomorphic behavior' of users towards technologies were demonstrated (Reeves and Nass 1996), various terms and concepts have been employed to denote this behavior and describe it. Paiva (2004) talks about empathy, Hoorn and Konijn (2003) address the concept of engagement, involvement, sympathy and their contrary, distance. Cassell and Bickmore (2003) adopt the Svennevig's theory of interpersonal relations.

We refer to Scherer's concept of interpersonal stance as a category which is "characteristic of an affective style that spontaneously develops or is strategically employed in the interaction with a person or a group of persons, coloring the interpersonal exchange in this situation (e.g. being polite, distant cold, warm, supportive, contemptuous)".

In particular, in referring to the social response of users to ECAs, we distinguish warm from cold social attitude, according to the Andersen and Guerrero's definition of interpersonal warmth (Andersen and Guerrero 1998) as "the pleasant, contented, intimate feeling that occurs during positive interactions with friends, family, colleagues and romantic partners".

We studied this attitude and the factors affecting it (Novielli et al., 2010) by observing the verbal and prosodic behavior of 60 subjects interacting with an ECA in a Wizard of Oz simulation study (de Rosis et al. 2007). More details about the WoZ study may be found in (Clarizio et al. 2006). In particular, we defined a markup language (Table 1) for the user moves after carefully examining our corpus and considering suggestions from the studies about verbal expression of social attitude (Andersen and Guerrero 1998, Polhemus et al. 2001, Swan 2002). Dynamic recognition of these individual signs during the dialogue enables not only to estimate the overall social attitude value but it also allows the agent to adapt its dialogue plan accordingly: for example, if the user tends to talk about herself, in the following moves the ECA will use this information to provide more appropriate suggestions. The overall social attitude of the user will be inferred dynamically from the history of the signs recognized during the dialogue to adapt the ECA's language style, voice and facial expression.

Table 1 Linguistic signs of Social Attitude and their definition

Linguistic Signs of Social Attitude with definition
Friendly self-introduction: The subjects introduce themselves with a friendly attitude (e.g. by giving their name or by explaining the reasons why they are participating in the dialogue).
Colloquial style: The subject employs an informal language, dialect, proverbs
Talks about self: The subjects provide more personal information about themselves than requested by the agent.
Personal questions to the agent: The subject tries to know something about the agent's preferences, lifestyle etc., or to give it suggestions in the domain.
Humor and irony: The subjects make some kind of verbal joke in their move.
Positive or negative comments: The subjects comment the agent's behavior, experience, domain knowledge, etc.
Friendly farewell: This may consist in using a friendly farewell form or in asking to carry-on the dialogue.

Three PhD students labeled independently the corpus of WoZ dialogues with our markup language. According to the result of the annotation experiment we defined a set of linguistic cues that could be considered as salient (Lee et al. 2002) for every given of social attitude. These cues are organized into semantic categories. Every new user move is categorized as 'showing a particular sign of social attitude' if it includes some word sequences belonging to semantic categories which are defined as 'salient' for the considered sign (Novielli et al., 2010). Recognition of linguistic signs of social attitude is performed by using Bayesian classification and can be enriched with acoustic analysis of user move, as described in (de Rosis et al. 2007).

3.1 Dynamic Modeling of the User Attitude

The user modeling procedure integrates (i) language analysis for linguistic cues extraction and (ii) a dynamic belief network (DBN) which considers the context in which the move was uttered. DBNs (Jensen 2001), also called time-stamped models, are local belief networks (called time slices) expanded over time; time slices are connected through temporal links to constitute a full model. The method allows us to deal with uncertainty in the relationships among the variables involved in the social attitude estimation (Table 2). The DBN formalism is particularly suitable for representing situations which gradually evolve from a dialog step to the next one. We applied results of the corpus analysis to learn from the annotated data a model of the user's mental state (Carofiglio et al. 2005) which includes the dimensions of interest for dialog adaptation. In particular: in learning the temporal part of our DBNs, we took every single user move in the corpus as an independent observation and applied the K2 algorithm (Cooper and Herskovitz 1992); in learning the temporal link between the monitored variable *Satt* at two subsequent time instants, we took every dialog as an observation to measure the conditional probability that *Satt* takes a given value at time t, given its value at time t-1.

The DBN (Figure 2) is employed to infer how the social attitude of the user evolves during the dialog in relation to the dialog history. The social attitude is the hidden variable of our model, that is the variable we want to monitor, which depends on observable ones, such as the 'stable' characteristics of the users (their background and gender), the context in which the move was entered (previous agent move) and the linguistic features of the user move recognized by our Bayesian classifier (leaf nodes of our DBN). Intermediate variables represent the signs of social attitude listed in Table 1.

Links among variables describe the causal relationships among stable characteristics of the users and their behavior, via intermediate nodes. DBNs, as employed in this paper, are said to be 'strictly repetitive models'. This means that structure and parameters of individual time slices is identical and temporal links between two consecutive time slices are always the same. We use a special kind of strictly repetitive model in which the Markov property holds: the past has no impact on the future given the present. In our simulations, every time slice corresponds to a user move, the stable user characteristics do not change from time to time (this is why we omitted the nodes Back and Gend from the figure) and temporal links are established only between dynamic subject characteristics in two consecutive time slices.

Table 2 Variables of our model

Variable category	Variable Name	Label
Stable user characteristics	Background	Back
	Gender	Gend
Context	Last agent move type	Ctext
	User move type	Mtype
Monitored variable	User attitude towards the agent	Satt
Signs of social attitude	Familiar style	Fstyl
	Friendly self-introduction	Fsint
	Talks about self	Perin
	Question about the agent	Qagt
	Friendly farewell	F-Fw
	Comments (positive and negative)	Comm
Result of linguistic analysis	Cues of familiar style	Pfstyl
	Cues of friendly self-introduction	Pfsint
	Cues of talks about self	Pperin
	Cues of questions to the agent	Pqagt
	Cues of friendly farewell	Pffw
	Cues of comments	Pcomm

At the beginning of interaction, the model is initialized by assigning a value to the stable user characteristics (e.g. female user with background in Humanities). At every dialog step, knowledge about the context and evidence produced by linguistic analysis are entered and propagated in the network: the model revises the probabilities of the social attitude node. The new probabilities of the signs of social attitude and stage of change are used in formulating the next agent move, while the probability of the social attitude node supports revising high-level planning of the agent behavior.

We performed an evaluation of the model to examine how a variation in the threshold of the probability of the monitored variable (Satt) affects sensitivity and specificity of the model in recognizing this feature. For more details about model validation please refer to the study described in (Clarizio et al. 2006).

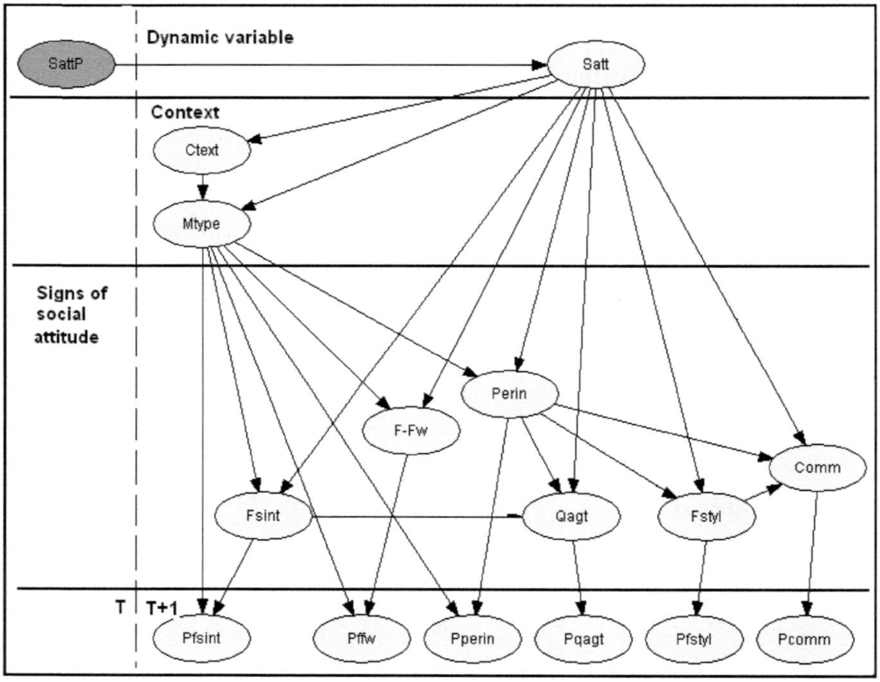

Fig. 2 User Model for the Social Attitude, a generic time-slice

4 A Model of Emotion Activation

In our emotion modeling method (de Rosis et al. 2003) we pay particular attention to how emotions change of intensity with time, how they mix up and how each of them prevails, in a given situation, according to the agent's personality and to the social context in which the dialog occurs. So far, we focused our attention on event-driven emotions in Ortony, Clore and Collin's (*OCC*) theory (Ortony et al. 1988). In this theory, *positive* emotions (happy-for, hope, joy, etc.) are activated by *desirable* events while *negative* emotions (sorry-for, fear, distress, etc.) arise after *undesirable* events.

Events concerning the agent are in the *Well-being* category (joy, distress), events concerning other people are in the *FortuneOfOthers* category (happy-for, sorry-for, envy and gloating) while future events are in the *Prospective* category (fear, hope). In Oatley and Johnson-Laird's theory, positive and negative emotions are activated (respectively) by the belief that some goal will be achieved or will be threatened (Oatley and Johnson-Laird 1987). A cognitive model of emotions that is built on this theory should represent the system of beliefs and goal behind emotion activation and endows the agent with the ability to *guess the reason why she feels a particular emotion and to justify it*. It includes the ingredients that enable representing *how the Agent's system of goals is revised* when emotions are felt and how this revision influences planning of subsequent dialog moves.

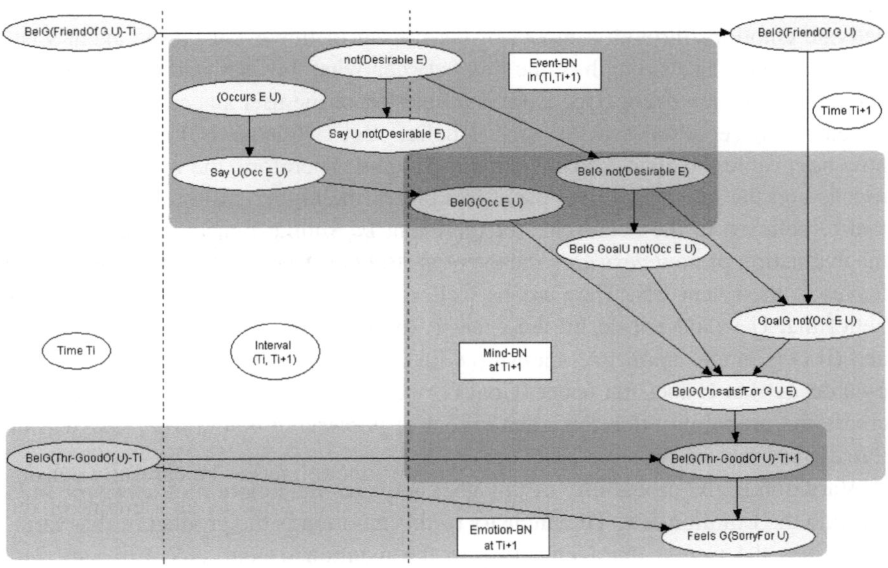

Fig. 3 A portion of the DBN that represents the agent's mental state showing the triggering of Sorry-For

Our model of emotion activation is represented with a *DBN* (Jensen 2001). We use *DBN*s as a goal monitoring system that employs the observation data in the time interval (T_i, T_{i+1}) to generate a probabilistic model of the agent's mind at time T_{i+1}, from the model that was built at time T_i. We employ this model to reason about the consequences of the observed event on the monitored goals. We calculate the intensity of emotions as a function of the *uncertainty* of the agent's beliefs that its goals will be achieved (or threatened) and of the *utility* assigned to achieving these goals. According to the utility theory, the two variables are combined to measure *the variation in the intensity of an emotion* as a product of the change in the probability to achieve a given goal, times the utility that achieving this goal takes to the agent (Carofiglio et al. 2008).

Let us consider, for instance, the triggering of *sorry-for* that is represented in Figure 3. This is a negative emotion and the goal that is involved, in this case, is *preserving others from bad*. The agent's belief about the probability that this goal will be threatened (Bel G (Thr-GoodOf U)) is influenced by her belief that some undesirable event E occurred to the user (BelG(Occ E U)). According to Elliott and Siegle (1993), the main variables influencing this probability are the desirability of the event (Bel G not(Desirable E)) and the probability that the agent attaches to the occurrence of this event (Bel G (Occ E U)). Other factors, such as the social context (Bel G FriendOf G U)), affect the emotion intensity. The model of the agent state at time T_{i+1} is built by automatically combining several *BN*s: the main one (Mind-BN) and one or more Event-BNs and Emotion-BNs. In the Event-BNs, the user moves are interpreted as *observable* consequences of occurred events that activate emotions through a model of the impact of this event on the agent's beliefs and goals. The strength of the link between what the user said (Say U (Occ E U)) and the *hidden* event (Occ E U) is a function of the user sincerity; the link between this observation and the agent's belief (Bel G (Occ E U)) is a function of how believable the agent considers the user to be. Therefore, the more sincere the user is and the more likely the event is a priori, the higher will be the probability that G believes in the occurrence of the event E. Similar considerations hold for the evaluation of how *desirable* the event is (Bel G (Desirable E)); these nodes are leaves of the Event-BN. They are, as well, roots of Mind-BN: they influence G's belief that U would not desire the event E to occur (Bel G Goal U ¬(Occ E U)) and (if G is in an *empathy* relationship with U and therefore adopts U's goals), its own desire that E does not occur (Goal G ¬(Occ E)). This way, they concur to increase the probability that the agent's goal of *preserving others from bad* will be threatened.

Variation in the probability of this goal activates the emotion of *sorry-for* in G through the Emotion-BN. The intensity of this emotion is the product of this variation times the *weight* the agent gives to the mentioned goal. According to Carbonell, we define a personality as a cognitively plausible combination of weights the agent gives to the goals represented in the model (Carbonell 1980).

The strength of the link between the goal-achievement (or threatening) nodes at two contiguous time instants defines the way the emotion, associated with that goal, decays, in absence of any event influencing it. By varying appropriately this strength, we simulate a more or less fast decay of emotion intensity. Different decays are attached to different emotion categories (positive vs. negative, FortuneOfOthers vs. Wellbeing and so on) and different temperaments are simulated, in which the *persistence* of emotions varies. The agents' affective state usually includes multiple emotions. Different emotions may coexist because an event produced several of them at the same time or because a new emotion is triggered while the previous ones did not yet decay completely. We describe in (Carofiglio et al. 2008) how we modeled the two mentioned mixing metaphors (*microwave oven* and *tub of water*, in Picard's terminology (Picard 1997)).

5 Dialog Modeling

The dialog manager includes three main layers:

1. a *Deliberative layer* that selects the goal with the highest priority and the correspondent plan and stores in the agenda the actions of the plan;
2. a *Communicative layer* that executes the next action in the agenda;
3. a *Reactive layer* that decides whether the goal priority should be revised, by applying reaction rules.

The dialog manager, and in particular the deliberative module, decides what goals to trigger and to pursue during the dialog, starting from the interpretation of the user move in terms of content and social attitude and according to the emotion triggered in the agent mind.

As the dialog evolves these factors may change what has been planned at two levels:

- by manipulating the inner aspects of the emotional response of our agent with an algorithm of activation/deactivation of its goals and of dynamic revision of their priorities;
- by deciding whether the agent should manifest its emotion and how.

Handling these issues is the main task of the Reactive Layer. The idea is that the agent has an initial list of goals, each with its priority, some of which are inactive: every goal is linked, by an application condition, to a plan that the agent can perform to achieve it. The communicative actions correspondent to active plans are put in the agenda maintained by the information state. The agent starts the dialog by executing these actions but, as we said in the Introduction, the agent applies some form reasoning on the user move. The recognized social attitude and the emotion triggered in the agent mind are used to implement social and emotion-based dynamic revision of goals and consequently of the dialog.

To achieve this aim, the following knowledge sources are employed by the dialog management modules:

1. *Agent's beliefs* that regard:
   - *long-term settings* that are stable during the dialog and influence the initial priority of the agent goals and therefore its initial plan, initiative handling and behavior: agent's personality, its role, its relationship with the user;
   - *short-term settings* that evolve during the dialog and influence goal priority change and plan revision: in particular, the emotional state of the agent and the social attitude of the user.

2. *Agent's goals* can be in one of the following relations among themselves:
   - *Priority:* $g_i < g_j$: g_i is more important, to the agent, than g_j. If this relation holds and no constraints or preconditions are violated by satisfying it, g_i will be achieved before g_j.

- *Hierarchy:* $H(g_i, (g_{i1}, g_{i2}, \ldots, g_{in}))$: the complex goal *gi* may be decomposed into simpler subgoals $g_{i1}, g_{i2},\ldots, g_{in}$, which contribute to achieving it.
- *Causal Relation:* $Cause(g_i, g_j)$: the plan achieving the source goal g_i is a precondition for the plan achieving the destination goal g_j.

3. *Plans* that are represented as context-adapted recipes; a recipe may be applied when some preconditions hold; its application affects the dialog state (agent's and user's mental state and interaction settings). In the healthy eating domain, our agent adopts the typical plan of intelligent advice systems:
   - *situation-assessment*, to acquire information about the user,
   - *describe-eating-problems*, to describe eating problems and their possible origin,
   - *suggest-solution*, to describe how to eat better and to overcome problems,
   - *persuade-to change*, to convince the users to change their eating habits.

4. *Reaction rules* that implement goal-revision strategies. They may produce, in general, the following effects on the dynamics of plan activation:
   - *add details* when the user asks for more information;
   - *reduce details* in case of urgency;
   - *abandon temporarily a plan* to activate a new subplan to reassure, motivate or provide more details;
   - *abandon a subplan* when its goal has been achieved: for example, when the user seems to know the information the agent is providing;
   - *substitute a generic subplan* with a more specific and situation-adapted one;
   - *revise the sequencing of plans,* to respond to the User request of taking the initiative. This is the most delicate situation: to be cooperative, the agent should leave aside its dialog plan and follow the user request; however, as we said, communicative goals may be linked by causal relations. Therefore, when the users show the intention to take the initiative in the dialog, the agent checks whether their goal may be activated immediately or whether some preconditions have first to be satisfied. It then satisfies these preconditions with the shortest subplan before satisfying the user request (De Carolis 1999).

As far as emotions and social factors are taken into account, according to Oatley and Johnson- Laird (1987) that claimed that *human plans are much more flexible than those so far explored in AI* , our reactive planning method takes these factors into account from two points of view:

1. *rules* regulating the *goal priority revision* by formalizing the following strategies:
   i. in case of *urgent events*, reduce the detail of information provided by upgrading the priority of "most relevant" subgoals and downgrading the priority of details;
   ii. in case of *desirable or undesirable events* occurred to the user, display *altruistic* social emotions (sorry-for and happy-for) by means of "full

expression" goals, that is by verbal and nonverbal means, and give them the highest priority; revise the priority of other goals; hide *egoistic* social emotions as envy and gloating;

iii. in case of *desirable events* occurred to the agent, activate surface expression goals: use verbal and nonverbal means to express them but leave the priority of other goals unvaried;

iv. in case of undesirable events (again occurred to the agent), activate behavior control goals: avoid displaying any emotional reaction by activating, at the same time, repair goals.

With these rules, we formalize a situation of empathic reaction in which the agent temporarily substitutes the presumed goals of the user for its own, when these goals are due to an emotional state of the user (Picard 1997). If an undesirable event occurs to the users, what they are presumed to need is to be convinced that the agent understands the situation and does its best to solve the problem. If something desirable occurs to them, they need to know that the agent shares their positive experience. If, on the contrary, the undesirable event does not concern the users, they probably want to be sure that this will not interfere negatively with the dialog.

2. *rules* regulating *emotion display*. Once felt, an emotion *E* may be hidden or displayed by the Agent *Ag*. Again, this decision (although not always taken consciously) may be influenced by personality factors and by the interaction context. In (De Carolis et al. 2001) we describe how this aspect of the emotional behavior of our agent can be modeled by means of rules that regulate activation of display goals.

   For example, the following rule:

```
If (Feel Ag E) ∧ Is-a (E WellBeing) ∧ (Valence E Nega-
tive) ∧ (Adoptive Ag U) then Goal G (¬ Display Ag E)
```

activates the goal *G* of hiding a socially negative emotion felt, because the agent has an adoptive relation with the user (i.e. fear, anger, etc.).

On the contrary, the following one:

```
If (Feel Ag E) ∧ Is-a (E WellBeing) ∧ (Valence E Posi-
tive) (Adoptive Ag U) then Goal G (Display Ag E)
```

activates the goal *G* of showing a positive emotion.

6 An Example of Dialog

To test how incorporating affective and social factors may contribute to the believability of the dialogue, we built a domain-independent simulator that uses the architecture described so far. The following (Figure 4) is an example dialogue in the application domain of healthy eating (Berry et al. 2005). In this dialogue the agent and user moves are denoted respectively with Vi and Uj. Vi moves are generated by our dialogue simulator, while Uj moves are introduced in free text.

V0: Hi, my name is Valentina! I'm here to suggest you how to eat well. What's your name?
U0: My name is Jana and I'm happy to interact with you. ***Friendly self-introduction***
V1: Good to hear your interest Jana! Hope you'll find our dialogue useful. How can I help you? ***ECA reciprocates the Colloquial attitude***
U1: I've been eating too much and badly in the past year and I don't feel comfortable with myself! ***Talk about self & Colloquial style***
V2: Oh! That's a bad situation indeed, but I understand! Let me try to see how can help you. Which is your age? ***ECA reciprocates the Warmth and shows a Sorry-for***
U2: I'm 17.
V3: Well: teens are notorious for being hungry frequently, because they need to eat more to support the major growth of their bodies. Maybe you only need an appropriate diet and change your eating habits.
U3: Yes, you are right! I definitely need a more regular regimen. ***Talk about self & Colloquial style***
V4: Good, let's try to do something then. Did you ever think to change your diet? ***ECA reciprocates the Warmth and shows a Happy-for***
U4: Yes, sometimes I did it and got very good results. Now, either because of my laziness or my stressing rhythms I can't lay myself into more regular habits. But now I want to try again! ***Talk about self & Colloquial style***
V5: Good, that's the way I want to see you! You seem to be in a more positive attitude now! ***ECA reciprocates the Friendly attitude & shows a Happy-for***

Fig. 4 An example of dialogue in the healthy eating domain

In the present prototype, translation of the user moves into a symbolic communicative act is quite primitive (just based on keyword analysis and on the dialogue context). For instance the U1 move 'I've been eating too much and badly in the past year and I don't feel comfortable with myself!' is interpreted as Say U (Occ EatingTooMuch U), Say U not(Desirable EatingTooMuch U). Symbolic communicative acts are inputs of the cognitive emotion model which, in this example, activates the Sorry-For. At the same time, linguistic cues of Colloquial Style and Talks about Self are detected and evidences about these signs contribute to increase the overall likelihood of observing a warm social attitude of the user. Hence, in the subsequent move (V2) the agent reacts by expressing her Sorry-For ('That's a bad situation indeed!') and by reciprocating the warm social attitude through the use of some small talk ('But I understand!').

The next move U2 does not show any particular sign of social attitude and does not provide any evidence which could potentially cause emotion triggering. Here the sorry-for decays due to the absence of any more stimuli. The dialogue goes on quite neutrally until the user claims her intention to change her diet, in U3. This event causes the triggering of a light Happy-For, whose intensity depends on the belief of the agent about the user sincerity, that is how true the agent beliefs the user wants to change her diet given that the user claimed it.

Then, the user reacts to the agent question by friendly talking about self. As a consequence, a higher level of the user social attitude is estimated, causing the agent to reply with a colloquial style in her next move ('Good, that's the way I want to see you!'). Moreover, the user states again her intention to change her diet causing an increase of the intensity of the Happy-For felt by the agent.

7 Exploiting the Potential of New Dialog Strategies

Providing an intelligent and conversational access to information, in our opinion, requires not only to dynamically adapt the information provision during the dialog but also to employ the most appropriate dialog strategy. An intelligent system should be able, for instance, to increase the user intention to accept the system suggestion/recommendation (Mahmood et al. 2009).

For instance, in the example dialogue (Figure 4), the persuasion attempt performed by the system is represented by a single dialogue move (V3), implementing the sub-plan 'persuade to change' in the scope of the overall dialogue plan of 'intelligent advice system'.

Though, users may rise objections or show perplexity during the dialog. Therefore the ECA has to reason in order to answer to the user reaction. For example, the following dialog excerpt (Figure 5) represents a variation to the one in Figure 4 in which the user objects to the ECA suggestion (U4).

The persuasive attempt in the dialogue in Figure 5 is generated by PORTIA (Mazzotta et al. 2007) a reasoning module able to decide on the most promising persuasion strategy to apply in a given scenario. It is a user-adapted persuasion module capable of simulating the persuasion process used by humans to convince someone to perform a given action. It mainly focuses on two typical aspects of the human persuasion in order to produce effective persuasion attempts in different contexts: on one hand, the ability of reasoning on the potential strength of alternative persuasive strategies for a given user, in order to select the most appropriate one; on the other hand, the capability of combining rational and emotional modes of persuasion. The system is based on the theory of *a-rational* persuasion (Miceli et al. 2006), and the strategies represented in the model are the result of a combination of theoretical (Walton 1992, 1996, Petty and Cacioppo 1896) and empirical (Mazzotta and de Rosis 2006) background. The key points of the system are the separation between *reasoning* and *argumentation* phases in the persuasion process (Walton 1990) and the use of *Belief Networks* to represent the uncertainty inherent in this form of practical reasoning (Pearl 1988).

V3: Well: teens are notorious for being hungry frequently, because they need to eat more to support the major growth of their bodies. Maybe you only need an appropriate diet and change your eating habits.
U3: How can I do?
V4: Why don't you try to eat more fruit and vegetables?
U4: I don't like them very much...
V5: Maybe you don't know that they have a lot of benefits on your health.
U5: I'm young and have a lot to do instead being worried for my health!
V6: Fruit and vegetables contribute to improve your appearance. FDA also says that they help you to have healthy skin and hair and it is an authoritative voice!
U6: But cooking vegetables is boring and I prefer spending my time among the people and making new friends.
V7: A dinner with fresh and tasty salads is easy to prepare and superb to spend good time with friends. I'm sure you can do it if you wish.
U7: Yes, you are right! I definitely need a more regular regimen.

Fig. 5 An example of the ECA persuasion attempt

The PORTIA's persuasion strategies (Mazzotta et al. 2008) are summarized in Table 3.

Table 3 A summary of the Persuasion Strategies used by PORTIA

<table>
<tr><td colspan="6">General induction of intentions strategy</td></tr>
<tr><td colspan="6">It may be summarized as follow: "If User has the goal g and he believes that doing the action a implies achieving g in a more or less near future, and he believes that has the ability to do a, then probably user intends to do a" (from Miceli et al. 2006).</td></tr>
<tr><td colspan="4">Rational induction of intention
It focuses on rational goals like 'to be in good health', 'to have a good appearance', …</td><td colspan="2">Emotional induction of intention
It focuses on rational goals like 'to make friends', to be in good mood', …</td></tr>
<tr><td colspan="6">Activation of goal strategy</td></tr>
<tr><td colspan="6">Activation through a belief or an emotion of an intermediate goal which is instrumental to the user's goal. It considers two possible applications: Rational Activation strategy or Emotional one.</td></tr>
<tr><td colspan="6">Induction of beliefs</td></tr>
<tr><td colspan="6">Argumentation about means-end implication. It represents the action-goal relation.</td></tr>
<tr><td>Appeal to Expert Opinion</td><td>Appeal to Popular Opinion</td><td>Appeal to Position to Know</td><td>Appeal to Friendly Personal Experience</td><td>Appeal to Examples</td><td>Others</td></tr>
</table>

In the *reasoning* phase, PORTIA exploits the information about the user in order to compute the degree of importance of the various -rational and emotional-goals, and infers the goals on which the persuasion strategy will focus. Using a "what-if" reasoning form it evaluates the persuasive power of different combination of strategies, and selects the most promising one, with respect to the goal of inducing in the user the intention to do a certain action.

In the *argumentation* phase, PORTIA constructs the arguments to express the strategy selected in the previous step by translating the output of the reasoning phase into a coherent discourse plan. The discourse plan is then translated by the dialogue manager into natural language messages used by the ECA as attempt to persuade the user.

8 Conclusions

The main goal of Intelligent Information Access is to provide a personalized and context-adapted access to information. In particular, systems implementing a conversational access to the information are enriched by exploiting human-computer interaction techniques (Mahmood et al. 2009). In this perspective, we present the architecture of an ECA which is able to exploit the knowledge conveyed by the user move in order to recognize her social attitude and goals and to behave 'believably' in its turn, by showing some forms of emotional intelligence. This research builds on prior work on affect modeling and dialog simulation. In particular, in this paper we combine social attitude and emotion modeling methods to build a scalable and open architecture for an emotionally and socially intelligent

ECA. In fact, each user move is rich of information (such as linguistic cues of social attitude) which goes beyond the pure content and meaning of sentences ('what user says'). These extra-rational information about the user state of mind can be exploited to enrich the user model and can be used by a socially and emotionally intelligent ECA, in order to tailor the dialogue strategy accordingly.

The two approaches to emotion and social attitude modeling have been validated in our previous research (Berry et al. 2005, de Rosis et al. 2007), with satisfying results.

We are aware of the limitations of our approach. In particular, translation of the user move meaning into symbolic form is currently implemented using a keyword-spotting based approach. In our future work, we plan to refine such analysis including contextual and acoustic information (Stolcke et al. 2000).

The main strength of the proposed ECA architecture is its openness and flexibility. In particular, we are able to simulate interactions in different conditions, by simply changing a few parameters describing the agent's personality. In this paper we show an example of adaptation by simulating the behavior of an empathic agent which reciprocates the social attitude of the user. In our future research we plan to perform evaluation studies in order to test which combination of personality traits of the agent best increases the user satisfaction. Moreover, we plan to investigate on the role that the interpersonal stances play in the display of emotions (De Carolis et al. 2001).

In our opinion, providing an intelligent and conversational access to information also requires the use of the most appropriate dialog strategy in order to increase the user intention to accept the system's suggestion. In this perspective, we describe an extension of our Agent's architecture with PORTIA, a module capable of simulating the persuasion process used by humans to convince someone to perform a given action.

Moreover, thanks to the independence of our architecture from the interaction mode, we plan to perform further investigation about spoken interaction. In particular, we will enrich the model for the analysis and interpretation of the user move using prosodic and acoustic parameters for improving the recognition of both (i) the actual communicative intention attached to the user move (De Carolis and Cozzolongo 2009) and (ii) the recognition of the user level of social attitude (de Rosis et al. 2007).

Acknowledgments. This research would not be without the encouragement, the suggestions and the teaching of Fiorella de Rosis. Her disappearance has been for us a sad loss, but her teaching incites us to continue the work.

References

Andersen, P.A., Guerrero, L.K.: Handbook of Communication and Emotions. In: Research, theory, applications and contexts. Academic Press, London (1998)

Berkovsky, S., Tsvi Kuflik, T., Ricci, F.: Cross-representation mediation of user models. User Modeling and User-Adapted Interaction 19(1-2), 35–63 (2009)

Berry, D.C., Butler, L.T., de Rosis, F.: Evaluating a realistic agent in an advice-giving task. International Journal of Man-Machine Studies 63(3), 304–327 (2005)

Bevacqua, E., Prepin, K., de Sevin, E., Niewiadomski, R., Pelachaud, C.: Reactive behaviors in SAIBA architecture. In: Workshop Towards a Standard Markup Language for Embodied Dialogue Acts, held in conjunction with AAMAS 2009 (2009)

Carbonell, J.C.: Towards a process model of human personality traits. Artificial Intelligence 15(1-2), 49–74 (1980)

Carofiglio, V., de Rosis, F., Novielli, N.: Dynamic User Modeling in Health Promotion Dialogs. In: Tao, J., Tan, T., Picard, R.W. (eds.) ACII 2005. LNCS, vol. 3784, pp. 723–730. Springer, Heidelberg (2005)

Carofiglio, V., de Rosis, F., Novielli, N.: Cognitive Emotion Modeling in Natural Language Communication. In: Tao, J., Tan, T. (eds.) Affective Information Processing, pp. 23–44. Springer, London (2008)

Carofiglio, V., de Rosis, F., Grassano, R.: Dynamic models of mixed emotion activation. In: Canamero, L., Aylett, R. (eds.) Animating expressive characters for social interactions, pp. 123–141. John Benjamins Publ. Co., Amsterdam (2008)

Cassell, J., Bickmore, T.: Negotiated collusion: modelling social language and its relationship effects in intelligent agents. User Modelling and User-Adapted Interaction 13(1-2), 89–132 (2003)

Clarizio, G., Mazzotta, I., Novielli, N., de Rosis, F.: Social Attitude Towards a Conversational Character. In: Proceedings of the 15th IEEE International Symposium on Robot and Human Interactive Communication, RO-MAN 2006, pp. 2–7 (2006)

Cooper, G.F., Herskovitz, E.: A Bayesian method for the induction of probabilistic networks from data. Machine Learning 9(4), 309–347 (1992)

De Carolis, B., Cozzolongo, G.: Interpretation of User's Feedback in Human-Robot Interaction. Journal of Physical Agents 3(2), 47–58 (2009)

De Carolis, B.: MyTutor: A Personal Tutoring Agent. In: Panayiotopoulos, T., et al. (eds.) IVA 2005. LNCS (LNAI), vol. 3661, pp. 478–488. Springer, Heidelberg (2005)

De Carolis, B., Pelachaud, C., Poggi, I., de Rosis, F.: Behavior Planning for a Reflexive Agent. In: Proceedings of the Seventeenth International Joint Conference on Artificial Intelligence, IJCAI 2001, pp. 1059–1066 (2001)

De Carolis, B., Pelachaud, C., Poggi, I., Steedman, M.: APML, a Mark-up Language for Believable Behavior Generation. In: Prendinger, H., Ishizuka, M. (eds.) Life-like Characters, Tools, Affective Functions and Applications, pp. 65–85. Springer, Heidelberg (2004)

De Carolis, B.: Generating Mixed-Initiative Hypertexts: A Reactive Approach. In: IUI, pp. 71–78. ACM, New York (1999)

de Rosis, F., De Carolis, B., Carofiglio, V., Pizzutilo, S.: Shallow and inner forms of emotional intelligence in advisory dialog simulation. In: Prendinger, H., Ishizuka, M. (eds.) Life-like Characters, Tools, Affective Functions and Applications, pp. 271–294 (2003)

de Rosis, F., Batliner, A., Novielli, N., Steidl, S.: 'You are Sooo Cool, Valentina!' Recognizing Social Attitude in Speech-Based Dialogues with an ECA. In: Paiva, A.C.R., Prada, R., Picard, R.W. (eds.) ACII 2007. LNCS, vol. 4738, pp. 179–190. Springer, Heidelberg (2007)

de Rosis, F., Castelfranchi, C., Goldie, P., Carofiglio, V.: Cognitive Evaluations And Intuitive Appraisals: Can Emotion Models Handle Them Both? In: Draft Chapter of the HUMAINE Handbook. Springer, Heidelberg (in press) de Rosis, F., Novielli, N., Carofiglio, V., Cavalluzzi, A., De Carolis, B.: User Modeling And Adaptation In Health Promotion Dialogs With An Animated Character. International Journal of Biomedical Informatics 39(5), 514–531 (2006)

de Rosis, F., Pizzutilo, S., Russo, A., Berry, D.C., Molina, F.J.N.: Modeling the User Knowledge by Belief Networks. User Model and User-Adapted Interaction 2(4), 367–388 (1992)

Elliott, C., Siegle, G.: Variables influencing the intensity of simulated affective states. In: Proceedings of the AAAI Spring Symposium on Mental States 1993, pp. 58–67 (1993)

Hoorn, J.F., Konijn, E.A.: Perceiving and Experiencing Fictitional Characters: An integrative account. Japanese Psychological Research 45(4), 250–268 (2003)

Huang, H., Nishida, T., Cerekovic, A., Pandzic, I.S., Nakano, Y.: The design of a generic framework for integrating ECA components. In: Proceedings of the 7th international Joint Conference on Autonomous Agents and Multiagent Systems, International Conference on Autonomous Agents. International Foundation for Autonomous Agents and Multiagent Systems, Richland, SC, Estoril, Portugal, May 12 - 16, vol. 1, pp. 128–135 (2008)

Jensen, F.V.: Bayesian Networks and Decision Graphs. Springer, Heidelberg (2001)

Kobsa, A.: User Modeling: Recent Work, Prospects and Hazards. In: Schneider-Hufschmidt, M., Kühme, T., Malinowski, U. (eds.) Adaptive User Interfaces: Principles and Practise. North Holland Elsevier, Amsterdam (1993)

Larsson, S., Traum, D.R.: Information state and dialogue management in the TRINDI dialogue move engine toolkit. Natural Language Engineering 6(3-4), 323–340 (2000)

Lee, C.M., Narayanan, S., Pieraccini, R.: Combining Acoustic and Language Information for Emotion Recognition. In: Proceedings of the seventh International Conference on Spoken Language Processing, ICSLP 2002 (2002)

Lee, E.J., Nass, C.: Effects of the form of representation and number of computer agents on conformity. In: Proceedings of Computer Human Interaction (CHI 1999), pp. 238–239 (1999)

Mahmood, T., Ricci, F., Venturini, A.: Learning Adaptive Reccomendation Strategies for Ondine Travel Planning. In: Information and Communication Technologies in Tourism 2009, pp. 149–160. Springer, Wien (2009)

Mazzotta, I., de Rosis, F.: Artifices for persuading to improve eating habits. In: AAAI Spring Symposium on 'Argumentation for consumers of health care', pp. 76–85 (2006); Technical Report SS-06-01

Mazzotta, I., de Rosis, F., Carofiglio, V.: PORTIA: a user-adapted persuasion system in the healthy eating domain. IEEE Intelligent Systems 22(6), 42–51 (2007)

Mazzotta, I., Silvestri, V., de Rosis, F.: Emotional And Non Emotional Persuasion Strength. In: Proceedings of AISB 2008, Symposium on 'Persuasive Technology', pp. 14–21 (2008)

Miceli, M., de Rosis, F., Poggi, I.: Emotional and non-emotional persuasion. Applied Artificial Intelligence: an International Journal 20(10), 849–880 (2006)

Neumann, B.C.: Scale in Distributed Systems. In: Casavant, T., Singhal, M. (eds.) Readings in Distributed Computing Systems, pp. 463–489. IEEE Computer Society Press, Los Alamitos (1994)

Novielli, N., de Rosis, F., Mazzotta, I.: User attitude towards an embodied conversational agent: Effects of the interaction mode. Journal of Pragmatics, doi:10.1016/j.pragma.2009.12.016 (2010)

Oatley, K., Johnson-Laird, P.N.: Towards a Cognitive Theory of Emotions. Cognition and Emotion 1, 29–50 (1987)

Ortony, A., Clore, G.L., Collins, A.: The cognitive structure of emotions. Cambridge University Press, Cambridge (1988)

Paiva, A. (ed.): Empathic Agents. Workshop in conjunction with AAMAS 2004 (2004)

Pearl, J.: Probabilistic Reasoning in Expert Systems: Networks of Plausible Reasoning. Morgan Kaufmann, San Mateo (1988)

Petty, R.E., Cacioppo, J.T.: The elaboration likelihood model of persuasion. In: Berkowitz, L. (ed.) Advances in experimental social psychology, vol. 19, pp. 123–205. Academic Press, New York (1986)

Picard, R.W.: Affective Computing. MIT Press, Cambridge (1997)

Poggi, I., Pelachaud, C., de Rosis, F.: Eye Communication in a Conversational 3D Synthetic Agent. AI Commun. 13(3), 169–182 (2000)

Polhemus, L., Shih, L.F., Swan, K.: Virtual interactivity: the representation of social presence in an on line discussion. In: Annual Meeting of the American Educational Research Association (2001)

Reeves, B., Nass, C.: The Media Equation: How people treat computers, television and new media like real people and places. Cambridge University Press, New York (1996)

Schröder, M. (2010) The SEMAINE API: Towards a Standards-Based Framework for Building Emotion-Oriented Systems. Advances in Human-Computer Interaction. Article ID 319406, 21 pages (2010) doi:10.1155/2010/319406

Stolcke, A., Coccaro, N., Bates, R., Taylor, P., Van Ess-Dykema, C., Ries, K., Shriberg, E., Jurafsky, D., Martin, R., Meteer, M.: Dialogue act modeling for automatic tagging and recognition of conversational speech. Computational Linguistics 26(3), 339–373 (2000)

Swan, K.: Immediacy, social presence and asynchronous discussion. In: Bourne, J., Moore, J.C. (eds.) Elements of quality online education, Nedham, MA. Sloan Center For Online Education, vol. 3, pp. 157–172 (2002)

Vilhjálmsson, H., Cantelmo, N., Cassell, J., Chafai, N.E., Kipp, M., Kopp, S., Mancini, M., Marsella, S., Marshall, A.N., Pelachaud, C., Ruttkay, Z., Thórisson, K.R., Welbergen, H., Werf, R.J.: The Behavior Markup Language: Recent Developments and Challenges. In: Pelachaud, C., Martin, J.-C., André, E., Chollet, G., Karpouzis, K., Pelé, D. (eds.) IVA 2007. LNCS (LNAI), vol. 4722, pp. 99–111. Springer, Heidelberg (2007)

Walton, D.N.: Argumentation Schemes for Presumptive Reasoning. N.J. Erlbaum, Mahwah (1996)

Walton, D.N.: The place of emotion in argument. The Pennsylvania State University Press (1992)

Walton, D.N.: What is reasoning? What is an argument? Journal of Philosophy 87, 399–419 (1990)

Annotating and Identifying Emotions in Text

Carlo Strapparava and Rada Mihalcea

Abstract. This paper focuses on the classification of emotions and polarity in news headlines and it is meant as an exploration of the connection between emotions and lexical semantics. We first describe the construction of the data set used in evaluation exercise "Affective Text" task at SEMEVAL 2007, annotated for six basic emotions: ANGER, DISGUST, FEAR, JOY, SADNESS and SURPRISE, and for POSITIVE and NEGATIVE polarity. We also briefly describe the participating systems and their results. Second, exploiting the same data set, we propose and evaluate several knowledge-based and corpus-based methods for the automatic identification of emotions in text.

1 Introduction

Emotions have been widely studied in psychology and behavior sciences, as they are an important element of human nature. They have also attracted the attention of researchers in computer science, especially in the field of human computer interaction, where studies have been carried out on facial expressions (e.g., [13]) or on the recognition of emotions through a variety of sensors (e.g., [35]).

Although only relatively little work has been carried out so far on the automatic identification of emotions in text [31, 1], the automatic detection of emotions in texts is becoming increasingly important from an applicative point of view. Consider for example the tasks of opinion mining and market analysis, affective computing, or natural language interfaces such as e-learning environments or educational/edutainment games.

Carlo Strapparava
FBK-IRST
e-mail: strappa@fbk.edu

Rada Mihalcea
University of North Texas
e-mail: rada@cs.unt.edu

G. Armano et al. (Eds.): Intelligent Information Access, SCI 301, pp. 21–38.
springerlink.com

For instance, the following represent examples of applicative scenarios in which affective analysis could make valuable and interesting contributions:

- *Sentiment Analysis.* tracking sentiment timelines in on-line forums and news [25, 5], review classification [43, 34], mining opinions from product reviews [20], etc., are examples of applications of these techniques. While positive/negative valence annotation is an active area in sentiment analysis, we believe that a fine-grained emotion annotation could increase the effectiveness of these applications.
- *Computer Assisted Creativity.* The automated generation of evaluative expressions with a bias on certain polarity orientation is a key component in automatic personalized advertisement and persuasive communication [8].
- *Verbal Expressivity in Human Computer Interaction.* Future human-computer interaction is expected to emphasize naturalness and effectiveness, and hence the integration of models of possibly many human cognitive capabilities, including affective analysis and generation. For example, the expression of emotions by synthetic characters (e.g., embodied conversational agents [11]) is now considered a key element for their believability. Affective words selection and understanding is crucial for realizing appropriate and expressive conversations [7].

This paper describes experiments concerned with the emotion analysis of news headlines. In Section 3, we describe the construction of a data set of news titles annotated for emotions, and we propose a methodology for fine-grained and coarse-grained evaluations. This data set was proposed at the "Affective Text" SEMEVAL task. Section 4 reports briefly the descriptions of the participating systems. In Section 5, we introduce several algorithms for the automatic classification of news headlines according to a given emotion. In particular we present several algorithms, ranging from simple heuristics (e.g., directly checking specific affective lexicons) to more refined algorithms (e.g., checking similarity in a latent semantic space in which explicit representations of emotions are built, and exploiting Naïve Bayes classifiers trained on emotion-annotated blogposts). Section 5.3 presents the evaluation of the algorithms and a comparison with the systems that participated in the SEMEVAL 2007 task on "Affective Text."

It is worth noting that the proposed methodologies are either completely unsupervised or, when supervision is used, the training data can be easily collected from online emotion-annotated materials such as blogs.

2 Background and Related Work

The characterization of emotions through linguistic analysis is a notoriously difficult task. On the one hand, emotions are not linguistic entities, and thus many of the previously proposed approaches for emotion detection were developed in a variety of other fields, including psychology, sociology, or philosophy. For instance, emotions have been studied with respect to facial expressions [13], action tendencies [17], physiological activity [4], or subjective experience [36].

On the other hand, one of the most convenient ways to access emotional content is through the use and analysis of language, and thus a number of previous efforts have been concentrated on the development of affective lexical resources.

One the first studies dealing with the referential structure of an affective lexicon is that of [29], consisting of an analysis of 500 words taken form the literature on emotions. Their goal was to develop a taxonomy of affective words, with special attention paid to the isolation of terms referring to emotions.

A well-known resource is General Inquirer [39]. The General Inquirer[1] is a mapping tool, which maps an input text file to counts in dictionary-supplied categories. The currently distributed version combines the "Harvard IV-4" dictionary content-analysis categories, the "Lasswell" dictionary content-analysis categories, and five categories based on the social cognition work of [38], for a total of 182 categories. Each category is a list of words and word senses. Currently, the category "negative" is the largest, with 2291 entries.

SentiWordNet[2] [14] is a lexical resource in which each synset s from WORDNET [16] is associated to three numerical scores $Obj(s)$, $Pos(s)$ and $Neg(s)$, indicating whether a synset term is objective, positive, or negative. The three scores are derived by combining the results produced by a committee of eight ternary classifiers.

The Affective Norms for English Words (ANEW) [9] provides a set of normative emotional ratings for a large number of words in the English language. This resource was built from analyses conducted on a wide variety of verbal judgments indicating the variance in emotional assessments along three major dimensions. The two main dimensions are "affective valence" (ranging from pleasant to unpleasant) and "arousal" (ranging from calm to excited). The third dimension is referred to as either "dominance" or "control."

Finally, WORDNET AFFECT[3] [41] is an extension of the WORDNET database that assesses a fine-grained emotion labeling of a subset of synsets suitable to represent affective concepts. In particular, one or more emotion labels (e.g. FEAR, JOY, LOVE) are assigned to a number of WORDNET synsets. There are also other labels for those concepts representing moods, situations eliciting emotions, or emotional responses. In this paper, we use WORDNET AFFECT in several of our experiments, as described in Section 5.

In addition to the task of emotion recognition and construction of affective lexical resources, related work was also concerned with opinion analysis and genre classification. Opinion analysis is a topic at the crossroads of text mining and computational linguistics, concerned with the identification of opinions (either positive or negative) expressed in a document [46, 44, 10, 33]. While opinion analysis deals with texts that are often affectively loaded, its focus is on subjectivity and polarity recognition, which is a coarser-grained level as compared to emotion recognition. Finally, related work in text genre classification was concerned with humor

[1] `http://www.wjh.harvard.edu/~inquirer/`

[2] `http://sentiwordnet.isti.cnr.it`

[3] WORDNET AFFECT is freely available for research purpose at `http://wndomains.itc.it`

recognition [27], male/female writing differences [22, 24], and happiness recognition in blogs [26].

3 Building a Data Set for Emotion Analysis

For the experiments reported in this paper we use the data set we developed for the SEMEVAL 2007 task on "Affective Text" [40].

The task was focused on the emotion classification of news headlines extracted from news web sites. Headlines typically consist of a few words and are often written by creative people with the intention to "provoke" emotions, and consequently to attract the readers' attention. These characteristics make this type of text particularly suitable for use in an automatic emotion recognition setting, as the affective/emotional features (if present) are guaranteed to appear in these short sentences.

The structure of the task was as follows:

Corpus: News titles, extracted from news web sites (such as Google news, CNN) and/or newspapers. In the case of web sites, we can easily collect a few thousand titles in a short amount of time.

Objective: Provided a predefined set of emotions (ANGER, DISGUST, FEAR, JOY, SADNESS, SURPRISE), classify the titles with the appropriate emotion label and/or with a valence indication (POSITIVE/NEGATIVE).

The emotion labeling and valence classification were seen as independent tasks, and thus a team was able to participate in one or both tasks. The task was carried out in an unsupervised setting, and consequently no training was provided. The reason behind this decision is that we wanted to emphasize the study of emotion lexical semantics, and avoid biasing the participants toward simple "text categorization" approaches. Nonetheless supervised systems were not precluded from participation, and in such cases the teams were allowed to create their own supervised training sets.

Participants were free to use any resources they wanted. We provided a set of words extracted from WORDNET AFFECT [41], relevant to the six emotions of interest. However, the use of this list was entirely optional.

3.1 Data Set

The data set consisted of news headlines drawn from major newspapers such as New York Times, CNN, and BBC News, as well as from the Google News search engine. We decided to focus our attention on headlines for two main reasons. First, news have typically a high load of emotional content, as they describe major national or worldwide events, and are written in a style meant to attract the attention of the readers. Second, the structure of headlines was appropriate for our goal of conducting sentence-level annotations of emotions.

Two data sets were made available: a development data set consisting of 250 annotated headlines, and a test data set consisting of 1,000 annotated headlines.[4]

3.2 Data Annotation

To perform the annotations, we developed a Web-based annotation interface that displayed one headline at a time, together with six slide bars for emotions and one slide bar for valence. The interval for the emotion annotations was set to $[0,100]$, where 0 means the emotion is missing from the given headline, and 100 represents maximum emotional load. The interval for the valence annotations was set to $[-100,100]$, where 0 represents a neutral headline, -100 represents a highly negative headline, and 100 corresponds to a highly positive headline.

Unlike previous annotations of sentiment or subjectivity [45, 32], which typically rely on binary $0/1$ annotations, we decided to use a finer-grained scale, hence allowing the annotators to select different degrees of emotional load.

The test data set was independently labeled by six annotators. The annotators were instructed to select the appropriate emotions for each headline based on the presence of words or phrases with emotional content, as well as the overall feeling invoked by the headline. Annotation examples were also provided, including examples of headlines bearing two or more emotions to illustrate the case where several emotions were jointly applicable. Finally, the annotators were encouraged to follow their "first intuition," and to use the full-range of the annotation scale bars.

The final annotation labels were created as the average of the six independent annotations, after normalizing the set of annotations provided by each annotator for each emotion to the 0-100 range. Table 1 shows three sample headlines in our data set, along with their final gold standard annotations.

Table 1 Sample headlines and manual annotations of emotions

	EMOTIONS						
	Anger	Disgust	Fear	Joy	Sadness	Surprise	Valence
Inter Milan set Serie A win record	2	0	0	50	0	9	50
Cisco sues Apple over iPhone name	48	8	10	0	11	19	-56
Planned cesareans not risk-free, group warns	0	0	61	0	15	11	-60

3.3 Inter-annotator Agreement

We conducted inter-tagger agreement studies for each of the six emotions. The agreement evaluations were carried out using the Pearson correlation measure, and are shown in Table 2. To measure the agreement among the six annotators, we first measured the agreement between each annotator and the average of the remaining five annotators, followed by an average over the six resulting agreement figures.

[4] The data set and more information about the task can be found at the SEMEVAL 2007 web site `http://nlp.cs.swarthmore.edu/semeval`

Table 2 Pearson correlation for inter-annotator agreement

EMOTIONS	
ANGER	49.55
DISGUST	44.51
FEAR	63.81
JOY	59.91
SADNESS	68.19
SURPRISE	36.07
VALENCE	
Valence	78.01

3.4 Fine-Grained and Coarse-Grained Evaluations

Provided a gold-standard data set with emotion annotations, we used both fine-grained and coarse-grained evaluation metrics for the evaluation of systems for automatic emotion annotation.

Fine-grained evaluations were conducted using the Pearson measure of correlation between the system scores and the gold standard scores, averaged over all the headlines in the data set.

We also ran coarse-grained evaluations, where each emotion was mapped to a 0/1 classification (0 = [0,50), 1 = [50,100]). For the coarse-grained evaluations, we calculated precision, recall, and F-measure.

4 Systems and Results Obtained in the AFFECTIVE TEXT Task

Five teams have participated in the "Affective Text" task as SEMEVAL, with five systems for valence classification and three systems for emotion labeling.

4.1 Participating Systems

The following represents a short description of the systems.

UPAR7:

This is a rule-based system [12] using a linguistic approach. A first pass through the data "uncapitalizes" common words in the news title. The system then used the Stanford syntactic parser on the modified title, and tried to identify what is being said about the main subject by exploiting the dependency graph obtained from the parser.

Each word was first rated separately for each emotion (the six emotions plus COMPASSION) and for its valence. Next, the main subject rating was boosted. Contrasts and accentuations between "good" or "bad" were detected, making it possible to identify surprising good or bad news. The system also takes into account: human will (as opposed to illness or natural disasters); negation and modals; high-tech context; celebrities.

The lexical resource used was a combination of SentiWordNet [15] and WORDNET AFFECT[41], which were semi-automatically enriched on the basis of the original trial data.

SICS:

The SICS team used a very simple approach for valence annotation based on a word-space model and a set of seed words [37]. The idea was to create two points in a high-dimensional word space - one representing positive valence, the other representing negative valence - and then projecting each headline into this space, choosing the valence whose point was closer to the headline.

The word space was produced from a lemmatized and stop list filtered version of the LA times corpus (consisting of documents from 1994, released for experimentation in the Cross Language Evaluation Forum (CLEF)) using documents as contexts and standard *tf.idf* weighting of frequencies. No dimensionality reduction was used, resulting in a 220,220-dimensional word space containing predominantly syntagmatic relations between words. Valence vectors were created in this space by summing the context vectors of a set of manually selected seed words (8 positive and 8 negative words).

For each headline in the test data, stop words and words with frequency above 10,000 in the LA times corpus were removed. The context vectors of the remaining words were then summed, and the cosine of the angles between the summed vector and each of the valence vectors were computed, and the headline was ascribed the valence value (computed as [cosine * 100 + 50]) of the closest valence vector (headlines that were closer to the negative valence vector were assigned a negative valence value). In 11 cases, a value of -0.0 was ascribed either because no words were left in the headline after frequency and stop word filtering, or because none of the remaining words occurred in the LA times corpus and thus did not have any context vector.

Table 3 System results for valence annotations

	Fine	Coarse			
	r	Acc.	Prec.	Rec.	F1
CLaC	**47.70**	**55.10**	**61.42**	9.20	16.00
UPAR7	36.96	55.00	57.54	8.78	15.24
SWAT	35.25	53.20	45.71	3.42	6.36
CLaC-NB	25.41	31.20	31.18	**66.38**	**42.43**
SICS	20.68	29.00	28.41	60.17	38.60

Table 4 System results for emotion annotations

	Fine	Coarse			
	r	Acc.	Prec.	Rec.	F1
		Anger			
SWAT	24.51	92.10	12.00	5.00	7.06
UA	23.20	86.40	12.74	**21.6**	**16.03**
UPAR7	**32.33**	**93.60**	**16.67**	1.66	3.02
		Disgust			
SWAT	**18.55**	97.20	0.00	0.00	-
UA	16.21	**97.30**	0.00	0.00	-
UPAR7	12.85	95.30	0.00	0.00	-
		Fear			
SWAT	32.52	84.80	25.00	14.40	18.27
UA	23.15	75.30	16.23	**26.27**	**20.06**
UPAR7	**44.92**	**87.90**	**33.33**	2.54	4.72
		Joy			
SWAT	**26.11**	80.60	35.41	**9.44**	**14.91**
UA	2.35	81.80	40.00	2.22	4.21
UPAR7	22.49	**82.20**	**54.54**	6.66	11.87
		Sadness			
SWAT	38.98	87.70	32.50	11.92	17.44
UA	12.28	88.90	25.00	0.91	1.76
UPAR7	**40.98**	**89.00**	**48.97**	**22.02**	**30.38**
		Surprise			
SWAT	11.82	**89.10**	11.86	10.93	11.78
UA	7.75	84.60	**13.70**	**16.56**	**15.00**
UPAR7	**16.71**	88.60	12.12	1.25	2.27

CLaC:

This team submitted two systems [3] to the competition: an unsupervised knowledge-based system (*CLaC*) and a supervised corpus-based system (*CLaC-NB*). Both systems were used for assigning positive/negative and neutral valence to headlines on the scale [-100,100].

CLaC:

The CLaC system relies on a knowledge-based domain-independent unsupervised approach to headline valence detection and scoring. The system uses three main kinds of knowledge: a list of sentiment-bearing words, a list of valence shifters and a set of rules that define the scope and the result of the combination of sentiment-bearing words and valence shifters. The unigrams used for sentence/headline classification were learned from WORDNETdictionary entries. In order to take advantage of the special properties of WORDNETglosses and relations, we developed a system that used the list of human-annotated adjectives from [19] as a seed list and learned additional unigrams from WordNetsynsets and glosses. The list was then expanded by adding to it all the words annotated with Positive or Negative tags in the General Inquirer. Each unigram in the resulting list had the degree of membership in the category of positive or negative sentiment assigned to it using the fuzzy net overlap score method described in the team's earlier work [2]. Only words with fuzzy membership score not equal to zero were retained in the list. The resulting list contained 10,809 sentiment-bearing words of different parts of speech.

The fuzzy net overlap score counts were complemented with the capability to discern and take into account some relevant elements of syntactic structure of the sentences. Two components were added to the system to enable this capability: (1) valence shifter handling rules and (2) parse tree analysis. The list of valence shifters was a combination of a list of common English negations and a subset of the list of automatically obtained words with increase/decrease semantics, complemented with manual annotation. The full list consists of 450 words and expressions. Each entry in the list of valence shifters has an action and scope associated with it, which are used by special handling rules that enable the system to identify such words and phrases in the text and take them into account in sentence sentiment determination. In order to correctly determine the scope of valence shifters in a sentence, the system used a parse tree analysis using MiniPar.

As a result of this processing, every headline received a system score assigned based on the combined fuzzy Net Overlap Score of its constituents. This score was then mapped into the [-100 to 100] scale as required by the task.

CLaC-NB:

In order to assess the performance of basic Machine Learning techniques on headlines, a second system ClaC-NB was also implemented. This system used a Naïve Bayes classifier in order to assign valence to headlines. It was trained on a small corpus composed of the development corpus of 250 headlines provided for this competition, plus an additional 200 headlines manually annotated and 400 positive and negative news sentences. The probabilities assigned by the classifier were mapped to the [-100, 100] scale as follows: all negative headlines received the score of -100, all positive were assigned the score of +100, and the neutral headlines obtained the score of 0.

UA:

In this system [23], in order to determine the kind and the amount of emotions in a headline, statistics were gathered from three different web Search Engines: MyWay, AlltheWeb and Yahoo. This information was used to observe the distribution of the nouns, the verbs, the adverbs and the adjectives extracted from the headline and the different emotions.

The emotion scores were obtained through Pointwise Mutual Information (PMI). First, the number of documents obtained from the three web search engines using a query that contains all the headline words and an emotion (the words occur in an independent proximity across the web documents) was divided by the number of documents containing only an emotion and the number of documents containing all the headline words. Second, associative score between a content word and an emotion was estimated and used to weight the final PMI score. The obtained results were normalized in the range 0-100.

SWAT:

SWAT [21] is a supervised system using an unigram model trained to annotate emotional content. Synonym expansion on the emotion label words was also performed, using the Roget Thesaurus. In addition to the development data provided by the task organizers, the SWAT team annotated an additional set of 1000 headlines, which was used for training.

4.2 Results

Tables 3 and 4 show the results obtained by the participating systems. The tables show both the fine-grained Pearson correlation measure and the coarse-grained accuracy, precision and recall figures.

The results indicate that the task of emotion annotation is difficult. Although the Pearson correlation for the inter-tagger agreement is not particularly high, the gap between the results obtained by the systems and the upper bound represented by the annotator agreement suggests that there is room for future improvements.

5 Automatic Emotion Analysis

In this section, we propose and evaluate several knowledge-based and corpus-based methods for the automatic identification of emotions in text, and compare the results with those obtained by the systems participating in the "Affective Text" task at SEMEVAL.

5.1 Knowledge-Based Emotion Annotation

We approach the task of emotion recognition by exploiting the use of words in a text, and in particular their co-occurrence with words that have explicit affective meaning. As suggested by Ortony et al. [30], we have to distinguish between words directly referring to emotional states (e.g., "fear", "cheerful") and those having only an indirect reference that depends on the context (e.g., words that indicate possible emotional causes such as "killer" or emotional responses such as "cry"). We call the former *direct affective words* and the latter *indirect affective words* [42].

As far as direct affective words are concerned, we follow the classification found in WORDNET AFFECT. This is an extension of the WORDNETdatabase [16], including a subset of synsets suitable to represent affective concepts. In particular, one or more affective labels (*a-labels*) are assigned to a number of WORDNETsynsets. There are also other a-labels for those concepts representing moods, situations eliciting emotions, or emotional responses. Starting with WORDNET AFFECT, we collected six lists of affective words by using the synsets labeled with the six emotions considered in our data set. Thus, as a baseline, we implemented a simple algorithm that checks the presence of this direct affective words in the headlines, and computes a score that reflects the frequency of the words in this affective lexicon in the text.

Table 5 Blogposts and mood annotations extracted from LiveJournal

Emotion	LiveJournal mood	Number of blogposts
ANGER	angry	951
DISGUST	disgusted	72
FEAR	scared	637
JOY	happy	4,856
SADNESS	sad	1,794
SURPRISE	surprised	451

A crucial aspect in the task of sentiment analysis is the availability of a mechanism for evaluating the semantic similarity among "generic" terms and affective lexical concepts. To this end we implemented a semantic similarity mechanism automatically acquired in an unsupervised way from a large corpus of texts (e.g., British National Corpus[5]). In particular we implemented a variation of Latent Semantic Analysis (LSA). LSA yields a vector space model that allows for a *homogeneous* representation (and hence comparison) of words, word sets, sentences and texts. For representing word sets and texts by means of an LSA vector, we used a variation of the *pseudo-document* methodology described in [6]. This variation takes into account also a *tf-idf* weighting schema (see [18] for more details). In practice, each document can be represented in the LSA space by summing up the normalized LSA vectors of all the terms contained in it. Thus a synset in WORDNET (and even all the words labeled with a particular emotion) can be represented in the LSA space, performing the pseudo-document technique on all the words contained in the synset. In the LSA space, an emotion can be represented at least in three ways: (i) the vector of the specific word denoting the emotion (e.g. "anger), (ii) the vector representing the synset of the emotion (e.g. {`anger`, `choler`, `ire`}), and (iii) the vector of all the words in the synsets labeled with the emotion. In this paper we performed experiments with all these three representations.

Regardless of how an emotion is represented in the LSA space, we can compute a similarity measure among (generic) terms in an input text and affective categories. For example in a LSA space built form the BNC, the noun "gift" is highly related to the emotional categories JOY and SURPRISE. In summary, the vectorial representation in the LSA allows us to represent, in a *uniform* way, emotional categories,

[5] BNC is a very large (over 100 million words) corpus of modern English, both spoken and written (see `http://www.hcu.ox.ac.uk/bnc/`). Other more specific corpora could also be considered, to obtain a more domain oriented similarity.

generic terms and concepts (synsets), and eventually full sentences. See [42] for more details.

5.2 Corpus-Based Emotion Annotation

In addition to the experiments based on WORDNET AFFECT, we have also conducted corpus-based experiments relying on blog entries from LiveJournal.com. We used a collection of blogposts annotated with moods that were mapped to the six emotions used in the classification. While every blog community practices a different genre of writing, LiveJournal.com blogs seem to more closely recount the goings-on of everyday life than any other blog community.

The indication of the mood is optional when posting on LiveJournal, therefore the mood-annotated posts we are using are likely to reflect the true mood of the blog authors, since they were explicitly specified without particular coercion from the interface. Our corpus consists of 8,761 blogposts, with the distribution over the six emotions shown in Table 5. This corpus is a subset of the corpus used in the experiments reported in [28].

Table 6 Sample blogposts labeled with moods corresponding to the six emotions

ANGER
I am so angry. Nicci can't get work off for the Used's show on the 30th, and we were stuck in traffic for almost 3 hours today, preventing us from seeing them. bastards
DISGUST
It's time to snap out of this. It's time to pull things together. This is ridiculous. I'm going nowhere. I'm doing nothing.
FEAR
He might have lung cancer. It's just a rumor...but it makes sense. is very depressed and that's just the beginning of things
JOY
This week has been the best week I've had since I can't remem ber when! I have been so hyper all week, it's been awesome!!!
SADNESS
Oh and a girl from my old school got run over and died the other day which is horrible, especially as it was a very small village school so everybody knew her.
SURPRISE
Small note: French men shake your hand as they say good morning to you. This is a little shocking to us fragile Americans, who are used to waving to each other in greeting.

In a pre-processing step, we removed all the SGML tags and kept only the body of the blogposts, which was then passed through a tokenizer. We also kept only blogposts with a length within a range comparable to the one of the headlines, i.e. 100-400 characters. The average length of the blogposts in the final corpus is 60 words / entry. Six sample entries are shown in Table 6.

The blogposts were then used to train a Naïve Bayes classifier, where for each emotion we used the blogs associated with it as positive examples, and the blogs associated with all the other five emotions as negative examples.

5.3 Evaluations and Results

We have implemented five different systems for emotion analysis by using the knowledge-based and corpus-based approaches described above.

1. WN-AFFECT PRESENCE, which is used as a baseline system, and which annotates the emotions in a text simply based on the presence of words from the WORDNET AFFECT lexicon.
2. LSA SINGLE WORD, which calculates the LSA similarity between the given text and each emotion, where an emotion is represented as the vector of the specific word denoting the emotion (e.g., JOY).
3. LSA EMOTION SYNSET, where in addition to the word denoting an emotion, its synonyms from the WORDNETsynset are also used.
4. LSA ALL EMOTION WORDS, which augments the previous set by adding the words in all the synsets labeled with a given emotion, as found in WORDNET AFFECT.
5. NB TRAINED ON BLOGS, which is a Naive Bayes classifier trained on the blog data annotated for emotions.

The five systems were evaluated on the data set of 1,000 newspaper headlines. As mentioned earlier, we conduct both fine-grained and coarse-grained evaluations. Table 7 shows the results obtained by each system for the annotation of the six emotions. The best results obtained according to each individual metric are marked in bold.

As expected, different systems have different strengths. The system based exclusively on the presence of words from the WORDNET AFFECT lexicon has the highest precision at the cost of low recall. Instead, the LSA system using all the emotion words has by far the largest recall, although the precision is significantly lower. In terms of performance for individual emotions, the system based on blogs gives the best results for JOY, which correlates with the size of the training data set (JOY had the largest number of blogposts). The blogs are also providing the best results for ANGER (which also had a relatively large number of blogposts). For all the other emotions, the best performance is obtained with the LSA models.

We also compare our results with those obtained by three systems participating in the SEMEVAL emotion annotation task: SWAT, UPAR7 and UA. Table 4 shows the results obtained by these systems on the same data set, using the same evaluation metrics.

For an overall comparison, we calculated the average over all six emotions for each system. Table 8 shows the overall results obtained by our five systems and by the three SEMEVAL systems. The best results in terms of fine-grained evaluations

Table 7 Performance of the proposed algorithms

	Fine	Coarse		
	r	Prec.	Rec.	F1
ANGER				
WN-AFFECT PRESENCE	12.08	**33.33**	3.33	6.06
LSA SINGLE WORD	8.32	6.28	63.33	11.43
LSA EMOTION SYNSET	17.80	7.29	86.67	13.45
LSA ALL EMOTION WORDS	5.77	6.20	**88.33**	11.58
NB TRAINED ON BLOGS	**19.78**	13.68	21.67	**16.77**
DISGUST				
WN-AFFECT PRESENCE	-1.59	0	0	-
LSA SINGLE WORD	**13.54**	**2.41**	70.59	**4.68**
LSA EMOTION SYNSET	7.41	1.53	64.71	3.00
LSA ALL EMOTION WORDS	8.25	1.98	**94.12**	3.87
NB TRAINED ON BLOGS	4.77	0	0	-
FEAR				
WN-AFFECT PRESENCE	24.86	**100.00**	1.69	3.33
LSA SINGLE WORD	**29.56**	12.93	**96.61**	**22.80**
LSA EMOTION SYNSET	18.11	12.44	94.92	22.00
LSA ALL EMOTION WORDS	10.28	12.55	86.44	21.91
NB TRAINED ON BLOGS	7.41	16.67	3.39	5.63
JOY				
WN-AFFECT PRESENCE	10.32	**50.00**	0.56	1.10
LSA SINGLE WORD	4.92	17.81	47.22	25.88
LSA EMOTION SYNSET	6.34	19.37	72.22	30.55
LSA ALL EMOTION WORDS	7.00	18.60	**90.00**	30.83
NB TRAINED ON BLOGS	**13.81**	22.71	59.44	**32.87**
SADNESS				
WN-AFFECT PRESENCE	8.56	**33.33**	3.67	6.61
LSA SINGLE WORD	8.13	13.13	55.05	21.20
LSA EMOTION SYNSET	13.27	14.35	58.71	**23.06**
LSA ALL EMOTION WORDS	10.71	11.69	**87.16**	20.61
NB TRAINED ON BLOGS	**16.01**	20.87	22.02	21.43
SURPRISE				
WN-AFFECT PRESENCE	3.06	**13.04**	4.68	6.90
LSA SINGLE WORD	9.71	6.73	67.19	12.23
LSA EMOTION SYNSET	12.07	7.23	89.06	13.38
LSA ALL EMOTION WORDS	**12.35**	7.62	**95.31**	**14.10**
NB TRAINED ON BLOGS	3.08	8.33	1.56	2.63

are obtained by the UPAR7 system, which is perhaps due to the deep syntactic analysis performed by this system. Our systems give however the best performance in terms of coarse-grained evaluations, with the WN-AFFECT PRESENCE providing the best precision, and the LSA ALL EMOTION WORDS leading to the highest recall and F-measure.

Table 8 Overall average results obtained by the five proposed systems and by the three SEMEVAL systems

	Fine	Coarse		
	r	Prec.	Rec.	F1
WN-AFFECT PRESENCE	9.54	**38.28**	1.54	4.00
LSA SINGLE WORD	12.36	9.88	66.72	16.37
LSA EMOTION SYNSET	12.50	9.20	77.71	13.38
LSA ALL EMOTION WORDS	9.06	9.77	**90.22**	**17.57**
NB TRAINED ON BLOGS	10.81	12.04	18.01	13.22
SWAT	25.41	19.46	8.61	11.57
UA	14.15	17.94	11.26	9.51
UPAR7	**28.38**	27.60	5.68	8.71

6 Conclusions

Affective computing deals with the automatic recognition and interpretation of emotions. While many studies have been carried out in the field of human-computer interaction, attempting to capture the user's physical state or behavior, only relatively little work has been carried out on the detection of emotions in texts. Written language is one of our main means of communication and, besides informative content, it also transmits attitudinal information, including emotional states. Thus, we believe that it is worthwhile to explore the task through existing state-of-the-art natural language processing techniques.

In this paper, we described the "Affective Text" task, presented at SEMEVAL-2007. The task focused on the classification of emotions in news headlines, and was meant as an exploration of the connection between emotions and lexical semantics.

After illustrating the data set, the rationale of the task, and a brief description of the participating systems, we presented several experiments in the automatic annotation of emotions in text. Through comparative evaluations of several knowledge-based and corpus-based methods carried out on the data set of 1,000 deadlines, we tried to identify the methods that work best for the annotation of emotions in text. The evaluation showed that different methods have different strengths, especially with respect to individual emotions. For instance, it seems that a machine learning classifier trained on blog data has good performance for recognizing JOY and ANGER, whereas a method based on semantic similarity is generally better for FEAR and SADNESS.

In future work, we plan to explore the lexical structure of emotions, and integrate deeper semantic processing of the text into the knowledge-based and corpus-based classification methods.

References

1. Aman, S., Szpakowicz, S.: Using roget's thesaurus for fine-grained emotion recognition. In: Proceedings of the International Joint Conference on Natural Language Processing, Hyderabad, India (2008)

2. Andreevskaia, A., Bergler, S.: Senses and sentiments: Sentiment tagging of adjectives at the meaning level. In: In: Proceedings of the 19th Canadian Conference on Artificial Intelligence, AI 2006. Quebec, Canada (2006)
3. Andreevskaia, A., Bergler, S.: CLaC and CLaC-NB: Knowledge-based and corpus-based approaches to sentiment tagging. In: Proceedings of SemEval-2007, Prague, Czech Republic (2007)
4. Ax, A.F.: The physiological differentiation between fear and anger in humans. Psychosomatic Medicine 15, 433–442 (1953)
5. Balog, K., Mishne, G., de Rijke, M.: Why are they excited? identifying and explaining spikes in blog mood levels. In: Proceedings of the 11th Meeting of the European Chapter of the Association for Computational Linguistics, EACL 2006 (2006)
6. Berry, M.: Large-scale sparse singular value computations. International Journal of Supercomputer Applications 6(1), 13–49 (1992)
7. Beskow, J., Cerrato, L., Granstrvm, B., House, D., Nordenberg, M., Nordstrand, M., Svanfeldt, G.: Expressive animated agents for affective dialogue systems. In: Proceedings of the Research Workshop on Affective Dialogue Systems, Kloster Irsee, Tyskland (2004)
8. Bozios, T., Lekakos, G., Skoularidou, V., Chorianopoulos, K.: Advance techniques for personalized advertising in a digital tv environment: the imedia system. In: eBusiness and eWork Conference, Venice, Italy, pp. 1025–1031 (2001)
9. Bradley, M., Lang, P.: Affective norms for english words (anew): Instruction manual and affective ratings. Tech. rep., The Center for Research in Psychophysiology, University of Florida (1999)
10. Breck, E., Choi, Y., Cardie, C.: Identifying expressions of opinion in context. In: Proceedings of IJCAI 2007, Hyderabad, India (2007)
11. Cassell, J.: Embodied conversational agents: Representation and intelligence in user interface. AI Magazine 22(3) (2001)
12. Chaumartin, F.: Upar7: A knowledge-based system for headline sentiment tagging. In: Proceedings of SemEval 2007, Prague, Czech Republic (2007)
13. Ekman, P.: Biological and cultural contributions to body and facial movement. In: Blacking, J. (ed.) Anthropology of the Body, pp. 34–84. Academic Press, London (1977)
14. Esuli, A., Sebastiani, F.: SentiWordNet: A publicly available lexical resource for opinion mining. In: Proceedings of the 5th Conference on Language Resources and Evaluation, Genova, IT (2006)
15. Esuli, A., Sebastiani, F.: SentiWordNet: A publicly available lexical resource for opinion mining. In: Proceedings of the Fifth International Conference on Language Resources and Evaluation (LREC 2006), Genoa, Italy (2006)
16. Fellbaum, C.: WordNet. An Electronic Lexical Database. The MIT Press, Cambridge (1998)
17. Frijda, N.: The Emotions (Studies in Emotion and Social Interaction). Cambridge University Press, New York (1982)
18. Gliozzo, A., Strapparava, C.: Domains kernels for text categorization. In: Proc. of the Ninth Conference on Computational Natural Language Learning (CoNLL 2005), Ann Arbor (2005)
19. Hatzivassiloglou, V., McKeown, K.: Predicting the semantic orientation of adjectives. In: Proceedings of the 35th Annual Meeting of the ACL, Madrid, Spain (1997)
20. Hu, M., Liu, B.: Mining and summarizing customer reviews. In: Proceedings of ACM SIGKDD Conference on Knowledge Discovery and Data Mining 2004 (KDD 2004), Seattle, Washington, pp. 168–177 (2004)

21. Katz, P., Singleton, M., Wicentowski, R.: SWAT-MP: The semeval-2007 systems for task 5 and task 14. In: Proceedings of SemEval-2007, Prague, Czech Republ. (2007)
22. Koppel, M., Argamon, S., Shimoni, A.: Automatically categorizing written texts by author gender. Literary and Linguistic Computing 4(17), 401–412 (2002)
23. Kozareva, Z., Navarro, B., Vazquez, S., Montoyo, A.: UA-ZBSA: A headline emotion classification through web information. In: Proceedings of SemEval-2007, Prague, Czech Republic (2007)
24. Liu, H., Mihalcea, R.: Of men, women, and computers: Data-driven gender modeling for improved user interfaces. In: International Conference on Weblogs and Social Media (2007)
25. Lloyd, L., Kechagias, D., Skiena, S.: Lydia: A system for large-scale news analysis. In: Consens, M.P., Navarro, G. (eds.) SPIRE 2005. LNCS, vol. 3772, pp. 161–166. Springer, Heidelberg (2005)
26. Mihalcea, R., Liu, H.: A corpus-based approach to finding happiness. In: Proceedings of the AAAI Spring Symposium on Computational Approaches to Weblogs, Stanford, CA, pp. 139–144 (2006)
27. Mihalcea, R., Strapparava, C.: Learning to laugh (automatically): Computational models for humor recognition. Computational Intelligence 22(2), 126–142 (2006)
28. Mishne, G.: Experiments with mood classification in blog posts. In: Proceedings of the 1st Workshop on Stylistic Analysis Of Text For Information Access (Style 2005), Brazile (2005)
29. Ortony, A., Clore, G., Foss, M.: The referential structure of the affective lexicon. Cognitive Science 11(3), 341–364 (1987)
30. Ortony, A., Clore, G.L., Foss, M.A.: The psychological foundations of the affective lexicon. Journal of Personality and Social Psychology 53, 751–766 (1987)
31. Ovesdotter, C., Roth, D., Sproat, R.: Emotions from text: machine learning for text-based emotion prediction. In: Proceedings of the conference on Human Language Technology and Empirical Methods in Natural Language Processing, pp. 579–586 (2005)
32. Pang, B., Lee, L.: A sentimental education: Sentiment analysis using subjectivity summarization based on minimum cuts. In: Proceedings of the 42nd Meeting of the Association for Computational Linguistics, Barcelona, Spain (2004)
33. Pang, B., Lee, L.: Opinion mining and sentiment analysis. Foundations and Trends in Information Retrieval 2(1-2), 1–135 (2008), http://dx.doi.org/10.1561/1500000011
34. Pang, B., Lee, L., Vaithyanathan, S.: Thumbs up? Sentiment classification using machine learning techniques. In: Proceedings of the Conference on Empirical Methods in Natural Language Processing (EMNLP 2002), Philadelphia, Pennsylvania, pp. 79–86 (2002)
35. Picard, R.: Affective computing. MIT Press, Cambridge (1997)
36. de Rivera, J.: A Structural Theory of the Emotions. International Universities Press, New York (1998)
37. Sahlgren, M., Karlgren, J., Eriksson, G.: SICS: Valence annotation based on seeds in word space. In: Proceedings of SemEval-2007, Prague, Czech Republic (2007)
38. Semin, G., Fiedler, K.: The cognitive functions of linguistic categories in describing persons: Social cognition and language. Journal of Personality and Social Psychology 54, 558–568 (1988)
39. Stone, P., Dunphy, D., Smith, M., Ogilvie, D.: The General Inquirer: A Computer Approach to Content Analysis. The MIT Press, Cambridge (1966)
40. Strapparava, C., Mihalcea, R.: SemEval-2007 task 14: Affective Text. In: Proceedings of SemEval 2007, Prague, Czech Republic (2007)

41. Strapparava, C., Valitutti, A.: WordNet-Affect: an affective extension of WordNet. In: Proc. of 4th International Conference on Language Resources and Evaluation, Lisbon (2004)
42. Strapparava, C., Valitutti, A., Stock, O.: The affective weight of lexicon. In: Proceedings of the Fifth International Conference on Language Resources and Evaluation, Genoa, Italy (2006)
43. Turney, P.: Thumbs up or thumbs down? semantic orientation applied to unsupervised classification of reviews. In: Proceedings of the 40th Annual Meeting of the Association for Computational Linguistics (ACL 2002), Philadelphia, pp. 417–424 (2002)
44. Wiebe, J., Mihalcea, R.: Word sense and subjectivity. In: Proceedings of 21st International Conference on Computational Linguistics, ACL 2006 (2006)
45. Wiebe, J., Wilson, T., Cardie, C.: Annotating expressions of opinions and emotions in language. Language Resources and Evaluation 39(2-3) (2005)
46. Wilson, T., Wiebe, J., Hwa, R.: Just how mad are you? Finding strong and weak opinion clauses. In: Proceedings of AAAI, pp. 761–769 (2004)

Improving Ranking by Respecting the Multidimensionality and Uncertainty of User Preferences

Bettina Berendt and Veit Köppen

Abstract. Rankings or ratings are popular methods for structuring large information sets in search engines, e-Commerce, e-Learning, etc. But do they produce the right rankings for their users? In this paper, we give an overview of major evaluation approaches for rankings as well as major challenges facing the use and usability of rankings. We point out the importance of an interdisciplinary perspective for a truly user-centric evaluation of rankings. We then focus on two central problems: the multidimensionality of the criteria that influence both users' and systems' rankings, and the randomness inherent in users' preferences. We propose multicriteria decision analysis and the integration of randomness into rankings as solution approaches to these problems. We close with an outlook on new challenges arising for ranking when systems address not only individuals, but also groups.

1 Introduction

Rankings or ratings are popular methods for structuring large information sets such as search engine results or products or documents in e-Commerce, e-Learning, and other environments. Their details are highly dependent on the intentions of suppliers and users. Ranking[1] is under research in the domain of concepts, cf. (Altman and Tennenholtz 2008; Marchant 2009; Rousseau 2008), and the domain of algorithms such as PageRank (Langville and Meyer 2006) and other algorithms for search engines (Chakrabarti 2003). However, there is often a lack of a comprehensive regard for the challenges of use and usability of systems (Berendt 2009).

Bettina Berendt
K.U. Leuven, Dept. of Computer Science, Leuven, Belgium
email: `http://www.cs.kuleuven.be/~berendt`

Veit Köppen
Otto-von-Guericke-Universität Magdeburg, Dept. of Technical & Business Information Systems, Magdeburg, Germany
email: `http://www.veit-koeppen.de/`

[1] Throughout this paper, we will regard ratings as a subclass of rankings.

G. Armano et al. (Eds.): Intelligent Information Access, SCI 301, pp. 39–56.
springerlink.com

The underlying idea of many current approaches is to reduce the inherent complexity and multidimensionality of the problem of finding the right ranking, and/or to hide this complexity from the user. The purpose of this paper is to draw attention to approaches that, instead, aim at showing complexity and making it manageable for users. We believe that such approaches, which give more control to users will ultimately fare better at empowering them than progressively simpler interfaces 'hiding' progressively opaque machine intelligence. In this paper, we will first, in Section 2, briefly review the simple model of evaluation, the challenges to it, and some solution proposals discussed in (Berendt 2009). We will add an important further challenge: users' 'mental rankings' may not be simple list or set structures. Instead, they may be only partially ordered structures, and they may exhibit random elements. In Section 3, we will then describe two classes of approaches for changing the search experience that address several of these challenges at once: multicriteria ranking and ranking with randomness. We derive these approaches from theoretical considerations. However, a deployment of these ideas could build on related ideas that are today already implemented in some search engines. We close with an outlook on future research.

2 Challenges for Evaluating Rankings

In this section, we present eight challenges that arise in the context of rankings. First, we introduce a simple model for a ranking system and the user of a ranking. For reasons of space, this section and its references are designed as exemplary rather than as a comprehensive survey.

2.1 A Simple Model of Systems, Users, and Evaluation

The system and the user have their respective ranking functions. A ranking is normally based on several criteria, and therefore the corresponding ranking function is based on a mapping f from $\mathbb{R}^m \to \mathbb{R}$, where m is the dimensionality of criteria. A ranking function $\Phi : A \to \mathbb{R}$ on the evaluated alternatives $\alpha \in A$ is then given by:

$$\Phi(\alpha) = f(v_1(x_1(\alpha)), v_2(x_2(\alpha)), \ldots, v_m(x_m(\alpha))), \tag{1}$$

where v_i evaluates a feature x_i of all the alternatives. With ranking function Φ, a ranking can be constructed. An assumption is that a weighting of the criteria is imposed by Φ. The characteristics of a ranking depend on this function.

A simple starting point for evaluation is to assume that (a) there is a system-generated ranking S, (b) there is a 'true' user-sided ranking U (usually not known to the user), and (c) evaluation consists of a goodness-of-fit test between S and U. There exist a multiplicity of measures for such a comparison, including precision and recall as well as various correlation measures, cf. (Herlocker et al. 2004; Vaughan 2004; Berendt 2009). Performance measure optimization is also under evolving research (Robertson and Zaragoza 2007). However, in this context the following challenges are not sufficient reflected.

2.2 Challenge 1: Context-Sensitive Tasks

Partial information often leads to inadequate decisions. This is of course also true in the domain of ranking. In the case of a Web search, the user does not provide all information that is required to obtain the best ranking. Reasons for this partial provisions may include cost, time, or the concern that overly restrictive queries may lead to zero results, and the consequent issuing of wider queries. Information is also partial on the other side of such a Web search. Not all information may be accessible, e.g., restrictions for robots, or analyzable, e.g., text within graphics. From the user's point of view, performing modified searches in further iterations may improve the results (see also Challenge 3 below). On the system side, new techniques, algorithms, and computational effort are required to improve the information status.

Examples of context properties that may lead to different user preferences are the geographic localization of users as well as linguistic and cultural factors or gender. Various measurements can indicate the values of these properties: IP address, language preference settings, or content choices, cf. for example (Mozilla Labs 2008; Liu and Mihalcea 2007; Hu et al. 2007; Kralisch and Köppen 2005; Kralisch and Berendt 2005; Berendt and Kralisch 2009).

2.3 Challenge 2: Purpose of Using the Ranking

A ranking of the user depends on the purpose of using the ranking. A classical example is a recommender system within a shopping portal that recommends items based on the currently viewed item. For example, users may be told that "people who bought this also bought ...", followed by a list of items sorted by strength of these associations. In this context, items that the user already possesses are not useful recommendations. Other misleading items might result from different categories like audio books instead of a DVD, when the use aims to find dance music.

Solution approaches that improve the ranking use new measures to produce non-obvious recommendations. *Novelty* brings elements into focus that the user is not familiar with. A formalization of novelty can be achieved by using interaction histories of the user or a quite homogeneous group of users, using publication dates to rank recent elements higher, e.g. (Herlocker et al. 2004). *Serendipity* ranks items higher that the user might not otherwise have discovered. Herlocker, Konstan, Terveen, and Riedl (2004) and Murakami, Mori, and Orihara (2008) give metrics for measuring serendipity, re-ranking items based on the probability of their usefulness for the current user as opposed to their usefulness for all users. An instance of this is the following measure proposed by Murakami, Mori, and Orihara (2008):

$$\textit{relative unexpectedness} = \frac{1}{N}\sum_{i=1}^{N} \textit{max}\left\{Pr\left(s_i\right) - Prim\left(s_i\right), 0\right\} \cdot \textit{isrel}\left(s_i\right) \cdot \frac{\textit{count}\left(i\right)}{i}, \quad (2)$$

where $Prim(s_i)$ is the result of a primitive prediction method for the ith item, $Pr(s_i)$ is the result of the used prediction method, and N is the number of items ranked. $isrel(s_1)$ is 1 if the item is related to the user's preferences, and 0 otherwise. $count(i)$

is the number of items suited to the user's preferences lying above the i-th rank in the recommendation list.

Zhang and Hurley (2009) address the problem of generating recommendation lists that not only contain novel and relevant items, but also exhibit *diversity*. They propose a method from economics for assessing inequality, based on concentration curves and an associated index, to analyse the bias of recommendation algorithms against the user's novel preferences.

2.4 Challenge 3: Rankings Are Results of Iterations

A ranking is often a result of iterations, where the query is adapted by the user to improve the returned item set and its ranking. An improvement for iterative obtained rankings is relevance feedback, cf. the survey in (Ruthven and Lalmas 2003). In each iteration, items are ranked according to their relevance to the user's query (by whichever relevance model is used). The basic idea is to reformulate the query by 'adding' features from items the user selected in the previous round, and 'deleting' features from the items the user did not select. Then, the items get re-ranked according to their relevance to the new query. Technically, a simple form is to compute a new query Q_t from the original query Q as follows:

$$Q_t = \alpha \cdot Q + \beta \frac{1}{|R_t|} \sum_{\forall x \in R_t} x + \gamma \cdot \frac{1}{|N_t|} \sum_{\forall y \in N_t} y, \qquad (3)$$

where R_t is the set of relevant items and N_t the set of non-relevant items.

A disadvantage of this simple approach is that all elements of a ranking have to be evaluated, otherwise non-observed elements are not treated adequately. However, this assumption is in general not realistic, as Section 2.5 will discuss.

The repetition of the same or similar queries by different users may be considered another form of iteration. Radlinski and Joachims (2005) propose to learn better rankings from the clickthrough behaviour of previous users with the same information need.

2.5 Challenge 4: Rankings Are Only Partially Perceived

The external ranking is often only observed partially. Research results on ranking attention, see for instance (Nielsen 2006; Eyetools 2008) support the hypothesis that a ranking is only perceived in the first positions. Whereas the first result is perceived by all users, only about 20 % of users look at the tenth element of the ranking (a generally-found pattern; the concrete numbers are from (Eyetools 2008)). Lewandowski and Höchstötter (2007) observe that this tendency of "information snacking" has increased over the past years. This may be explained by the increased trust of users in search engines' rankings, derived from past interactions with the search engine in which the highly ranked elements were indeed useful. It is also a consequence of well-known patterns in human reading, cf. the observation that newspaper stories "above the fold" are read most, help sell the paper, and therefore are most attractive for advertising.

To address this challenge, evaluation measures that weight the top positions more strongly have been proposed. They include the discounted cumulative gain or the half-life utility metric:

$$U = \frac{max\{r_i - d, 0\}}{2^{(i-a)/(a-1)}}, \tag{4}$$

where r_i is the user's rating of the item, d is the default rating, and a is the half-life: the rank of the item on the list such that there is a 50% chance that the user will view that item.

2.6 Challenge 5: Rankings Are Embedded in Information Systems

The system's ranking S and the user's preferences do not exist in a vacuum. Rather, the user and system form an information system in which IT tries to produce a good approximation of users' needs and present it appropriately. Therefore, the designers of IT systems have to find a usable solution that is accepted (and used) by the users. This is a main requirement for the success of Internet search engines. The search engine usability affects in this context effectiveness, efficiency, and satisfaction.

Usability is defined as the effectiveness, efficiency, and satisfaction with which specified users achieve specified goals in particular environments (Dix et al. 1998), which is captured by ISO 9241 (International Organization for Standardization 2007; International Organization for Standardization 2008). *Effectiveness* is the accuracy and completeness with which specified users can achieve specified goals in particular environments. *Efficiency* denotes the resources expended in relation to the accuracy and completeness of goals achieved. *Satisfaction* are the comfort and acceptability of the working system. A breakdown by "usability objectives" is helpful for operationalising effectiveness, efficiency and satisfaction metrics (Dix et al. 1998). A comprehensive treatment of usability metrics is given by Tullis and Albert (2008).

For adaptive systems, further usability goals have to be taken into account, cf. (Jameson 2003). These goals include that the system should leave the user in control. Another important aspect is the transparency of the systems actions, so the actions are on the one hand understandable and on the other hand predictable for the user. Privacy issues have to be respected as well. Last but not least, the system should not limit the users' breadth of experience.

2.7 Challenge 6: Framing Influences User Preferences

Challenge 4 emphasized that the layout of information on the screen may have important influences on whether this information is perceived. However, visual perception is not the only source of bias; the way in which information is phrased is another key determinant of how information is understood – and how preferences are formed.

"Framing" is an inevitable process of selective influence over an individual's perception of the meanings attributed to words or phrases. A frame defines the packaging of a certain piece of content such that certain interpretations are encouraged and others are discouraged. A well-known demonstration of this is due to Tversky and Kahneman (1981), who gave experiment participants two versions of the same facts. The task was to decide between two medical treatments for an epidemic. The "positive frame" described option A as "saving 200 people" and option B as "a one-third probability that 600 people will be saved and a two-thirds probability that no people will be saved". The "negative frame" described option C as "400 people dying" and option D as "a one-third probability that nobody will die and a two-thirds probability that 600 people will die". Even though probabilistically, A and C as well as B and D are equal, people consistently preferred A over B in the positive frame and D over C in the negative frame.

Framing effects also occur in Web searches and recommendations. Framing the quest for highly personal information in terms of a gain in personalization makes irrelevant questions seem relevant and non-legitimate questions seem legitimate, cf. (Berendt et al. 2005). Framing a ranked recommendation list as arising from data mining creates the impression of personalization (even if an identical list is presented to everyone), and it leads to a higher willingness to disclose personal information (Kobsa and Teltzrow 2005).

With regard to document search, framing effects pose a specific challenge: If ranking is based only on the "factual content" of documents, it will miss out on an important source of human preferences of documents over others – recall that in the example, the "factual content" of options A and C was identical, as was that of B and D, but A was preferred over C and D over B. If, on the other hand, document-processing techniques such as natural-language understanding were able to extract framing, how should the system deal with it when producing the ranking? Should it try to comply with the user's favourite framing or framing-induced preferences? Or should it try to counteract them, for example by focussing on presenting a diversity of framings? We believe that successful solution approaches should, first and foremost, help people become more aware of the presence of framing and its influences.

Taking a wider perspective, one can observe that the habit of using search engines and other rankings also constitutes a certain frame. It can be expected that this will also influence people's perceptions of rankings. We will investigate this question in the following challenge.

2.8 *Challenge 7: System-Use Dynamics and Erroneous Beliefs About Algorithms*

Information retrieval and Web search have become such an integral part of everyday activities that there is a strong tendency to (blindly) regard search-engine results as "the truth". To find out whether users relied more on a trusted system's ranking or on their own assessments of the result snippets, Pan, Hembrooke, Joachims, Lorigo, Gay, and Granka (2007) exposed people to one of three experimental conditions: Upon entering a search query, users received either the ranking returned by Google, this ranking with the first and second results exchanged, or this ranking in reversed order. In all three conditions, the interface was the familiar Google result list layout. In all three conditions, users displayed the typical behaviour of viewing mainly the first two or three results (see Section 2.5), together with similarly strong preferences in clickthrough behaviour. Although users in the "reversed condition" did appear to notice something unusual (as witnessed by longer viewing times of the results page), they attributed the problems to themselves, stating that they "did not have much luck with several of the questions" or that they "could not think of the right search terms".

This may be regarded as evidence of an exaggerated trust in numbers, fed by a reliance on the "brand" of well-known search engines (Jansen et al. 2007) and by erroneous or naïve beliefs in poorly understood algorithms. The latter phenomenon has been well-studied in another area in which rankings are ubiquitous and used and (often mis-)interpreted especially by non-experts: bibliometrics, see for example (Glänzel 2008).

2.9 Challenge 8: User Rankings Are Not Simple

Throughout the previous sections, we have assumed that the user's ranking is a simple list, i.e. a total order. However, research suggests that preference orders such as those described in the economic theory of households (Varian 2007) appear much more often. These orders are only partial. Furthermore, user assessments are often better described by random variables than by fixed rankings.

Solving the conflict of partial orders might be done by using similarity measures that respect partial orders, e.g., corrected pairs as stated by Sørensen, Lerche, and Thomsen (2006). However, a restriction to such changes in measures disregards the other challenges discussed in this paper. We will therefore proceed to discuss partial orders and randomness in connection with these challenges, in particular the multiple dimensions influencing ranking and the need to take an integrated view of the information system in which users, systems, and their rankings interact with each other.

3 Making Multidimensionality Transparent

The "right" ranking is determined by multiple dimensions: text relevance to a query, novelty, geographic proximity of search results, etc. This observation is captured in the basic formulation of ranking functions Φ (see Equation 1), in the challenges posed by context and purpose (see Sections 2.2 and 2.3), and it may become manifest in the queries issued and results selected in different iterations of a search episode (see Section 2.4).

Not only do multiple *dimensions* exist, they also have different *weights*. For example, a simple version of Φ could be a weighted average of dimension measures. These weights may differ between people, situations, etc. In the following, we will also use *criteria* as another term to denote dimensions, in order to emphasize that they are criteria for choosing or ranking one alternative over others.

In Section 2, we have concentrated on solution approaches for multidimensionality that focus on helping systems derive the right dimension weights automatically. However, it may well be that a more transparent and user-led control of weights will involve users more strongly and lead to greater satisfaction with the software (see usability criterion "control" in Section 2.6) and/or even lead to better-matching ranking results. It remains to be investigated whether usability/ranking quality is a trade-off and if so, how it is assessed by users. It will certainly make the workings of algorithms more transparent and understandable, thus hopefully reducing blind and erroneous beliefs in algorithms (see Section 2.8).

Special challenges arise when users' multiple criteria are to be taken into account: How can bounds on rationality such as merely partial preference orders be taken into account (see Section 2.9)? How can interface design create the best affordances for inspecting and setting criteria and weights (see Sections 2.5 and 2.6)? In addition, new questions arise, such as: Can a user-defined criteria setting re-frame the search results or their perception by the search-engine user (see Section 2.7)?

3.1 Multidimensional Ranking

To deal with a multitude of criteria, a reduction of dimensionality is nearly always required. This is performed by function Φ, see Eq. 1. This loss of dimensionality might lead to an information loss. Therefore, we present an approach that modifies f via an intermediate mapping $\mathbb{R}^m \to \mathbb{R}^n$, where $m \geq n$. (Φ remains a mapping from A into $\mathbb{R}$.) The resulting n criteria are used to build a set of rankings, where only one criterion is changed at a time. Note that these n criteria may be aggregated indicators, resulting from transformations of the original m criteria.

The presentation of the top alternatives within a visualization as in Fig. 1 enhances the understanding from the user's viewpoint. The criteria of this example are derived from the observations described in Sections 2.0, 2.2 and 2.6, and their values are shown together with three fictitious alternatives $\alpha_{[1|2|3]}$ to be ranked. Each criterion is assigned to one spoke of the chart, and for each alternative to be ranked, a line connecting its values on the criteria is drawn. Thus, the chart is a concise way of showing the rankings of all alternatives along all criteria simultaneously. This control of dimensionality with a Kiviat chart is based on the Balanced Scorecard approach proposed and extended by Kaplan and Norton (1996, 2004). The n criteria may be clustered into groups, where different groups are possible. One possible grouping of the criteria of Fig. 1 will be shown below in Fig.3. Navigation through these rankings is necessary due to complexity, and it is possible with visualization techniques. The Kiviat chart is an appropriate choice for such a representation.[2]

The explicit display of different criteria for navigation between alternatives is used in several real-life search tools. It is popular in e-Commerce and other

[2] Parallel coordinates (d'Ocagne 1885; Inselberg 1985) are information-equivalent to these charts, but they lack the "holistic Gestalt" induced by the circular spoke arrangement of the axes.

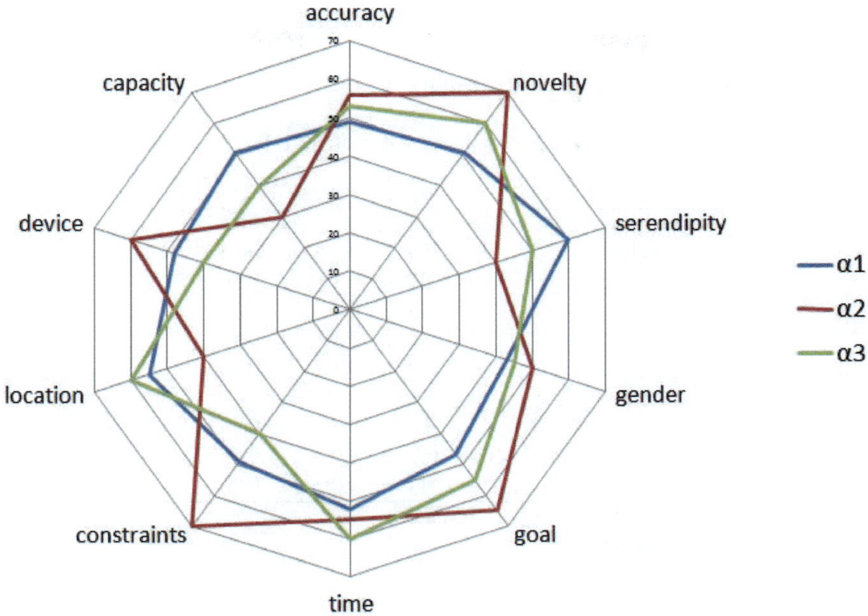

Fig. 1 Kiviat Chart of Selected Ranking Criteria

sites that allow users to order the found items by price, popularity, or other criteria; an interaction technique known as *faceted search*, cf. (Hearst 2006). Berendt and Trümper (2009) define several dimensions of document similarity (textual similarity, named-entity similarity, date similarity, and source class) and provide a visualisation of document sets in this space. Documents are arranged in a display plane such that they are visually ordered, along the different criteria, by their similarity to a focus document.

An obvious limitation of a Kiviat chart or other representations that show individual criteria rankings is that often, no alternative will dominate all the others on every criterion. Thus, users may prefer to see a ranking that presents *one* (preferably total) order on the alternatives, summarizing their rankings along the different criteria. This is the essence of the most commonly used Φ functions, of which a very simple one will be described next.

In Table 1, we extend the example by using a simple Φ function. The table contains fictitious values $v(x(\alpha))$ that evaluate the three alternatives α from Fig. 1. We also present two different sets of weights for these criteria and, for each set of weights, a summarising one-dimensional score and the ranking derived from it. The score is the inner product of the criteria values vector and the weights vector.

Depending on the weights, the obtained ranking differs. For the first set of weights, the ranking is: alternative 2 is followed by 3 and 1. For the second set of weights, where only small changes in the weights are used, the ranking changes completely. This change is not obvious or traceable for the user if only the ranking is provided – this constitutes the information loss. This may be problematic when the weights are uncertain, and the presentation of a result (based on whichever set of weights) creates a sense of false certainty.

Table 1 Evaluation of Three Alternatives $\alpha 1$, $\alpha 2$, $\alpha 3$

Criterion	$\alpha 1$	$\alpha 2$	$\alpha 3$	*Weights 1*	*Weights 2*
accuracy	49	56	53	0.10	0.05
novelty	51	70	60	0.10	0.05
serendipity	60	40	50	0.10	0.10
gender	43	50	45	0.10	0.10
goal	47	65	55	0.10	0.10
time	52	55	60	0.10	0.10
constraints	49	70	40	0.10	0.10
location	55	40	60	0.20	0.15
device	48	60	40	0.05	0.05
capacity	51	30	40	0.05	0.20
Score 1	51.05	53.10	52.30		
Ranking 1	3	1	2		
Score 2	50.95	49.30	49.65		
Ranking 2	1	3	2		

While this is a simple and clear procedure for aggregating different criteria into a ranking, it raises the question of how to acquire those weights. In the following two sections, we will give a brief overview of proposals for addressing these questions.

3.2 Weighting the Criteria: Multicriteria Decision Analysis

An improvement of the system's ranking can be obtained if the function Φ of the system is close the the user's one. However, a complete specification is usually not possible due to the complexity and the change over context and time. Therefore, Φ should be adjusted to make it robust with respect to small temporal changes, as well as lower-dimensional and thus manageable. In a given set of characteristics stated by the system, two possible scenarios are: on the one hand the user might specify an order for each feature, on the other hand the weights of each feature are revealed. Whereas the first possibility is too complex due to the fact that a generation of rankings for each feature is required, the second possibility is much more prudent. In the following approaches are described to obtain these weights.

In this section we use Multi Criteria Decision Analysis (MCDA) to obtain a ranking where all criteria are respected. MCDA is used to obtain a ranking of alternatives in a decision process. However, it can also be applied for obtaining a ranking in the context of Internet search engines where the decision has to be taken, which of all available elements fits best to a certain information need or query.

In the domain of MCDA, many methods can be differentiated. In Fig. 2 the main approaches are presented according to the classification by Schneeweiß (1991). Further information is given in (Figueira et al. 2005; Bamberg and Coenenberg 2002; Schneeweiß 1991).

A first differentiation can be made on the top level. Methods "with preference functional" are those for which at least a weak preference order of alternatives is available or required. The other approaches do not need such an order. In practice, methods without a preference functional are used when a transitive order of alternatives cannot be found.

A further classification of the preference order functional is also depicted in Fig. 2. When substitution rules between alternatives are given, a "preference function" is available. All alternatives have to be assigned a measurement, and these measurements must be interval-scaled. A "preference index" is used if a weak order on alternatives is given. All alternatives might be ordered directly or indirectly via an order on attributes. The distinct evaluation of utility function and weights has to be combined afterwards again to obtain a holistic result. When only a "partial preference functional", i.e. only a partial preference order, is given, iterations of the ranking process are useful. In each iteration, another order relationship is elicited, and this information is used to improve the ranking. However, if a ranking does not differentiate the alternatives sufficiently, a further iteration is required. In all these approaches, a characteristic of an attribute can be compensated by one or more others. A "non-compensating preference functional" orders all alternatives according to the importance of criteria. Ordering proceeds by considering the most important criterion first and then recursively employing this procedure on all the other criteria. An example is lexicographic ordering.

Methods "without a preference functional", also called outranking, assume that an ordering is not possible or misleading. These approaches are used to find the top solution or a classification of alternatives.

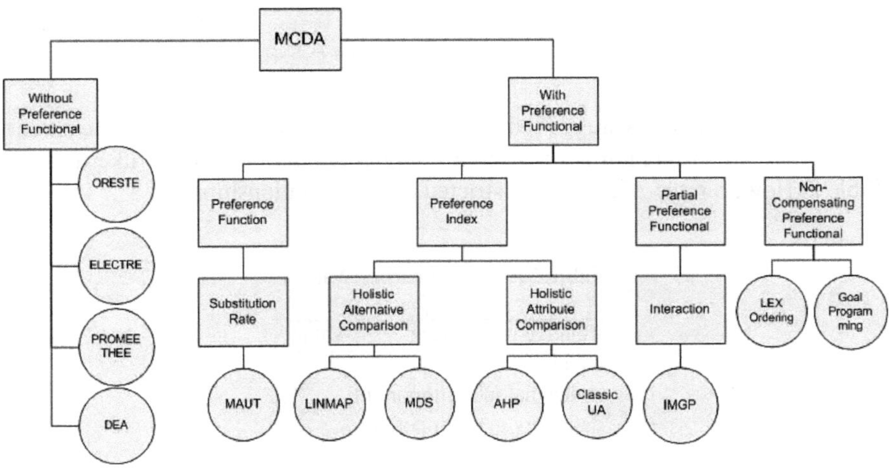

Fig. 2 MCDA Classification, adapted from (Schneeweiß 1991)

In this paper, we restrict ourselves to the Analytic Hierarchy Process as one method with a preference functional, more specifically a preference index. AHP is a method based on "holistic attribute comparison". This means that attributes as a whole are compared to one another irrespective of these attributes' individual values (and possible fine-grained preferences on them).

3.3 The Analytic Hierarchy Process Method

The Analytic Hierarchy Process (AHP) (Saaty 1980) is a special case of utility value analysis (Keeney and Raiffa 1976). The weights are holistic and obtained independently of the evaluation function by using pairwise comparisons between attributes. This makes it possible that even if the preference order is inconsistent (non-transitive), the evaluation function is in a consistent state. Therefore, this approach is promising for building rankings with partial orders. However, an important disadvantage of AHP is that when new alternatives are integrated, large and seemingly unmotivated re-rankings can occur. In addition, if the criteria space is high-dimensional, it is costly to obtain all pairwise comparisons.

In the first phase of the AHP, all required criteria are selected and classified into a hierarchy. This classification is used to reduce the pairwise comparisons. A comparison is only necessary for attributes on the same level and with the same parent. In Fig. 3, we present a possible classification for a selection of the criteria introduced above in Fig. 1.

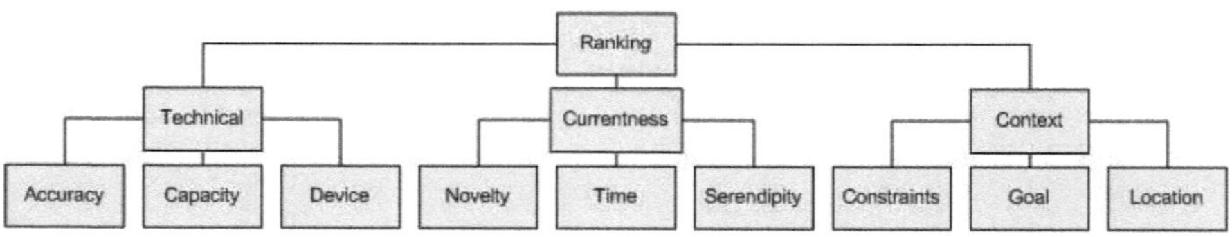

Fig. 3 A Hierarchy for Ranking Criteria

Pairwise comparisons are carried out at each level for nodes with the same parent. In practice, the evaluation is often done using comparison relations like those in Table 2. However, the AHP is not restricted to these relationships.

Table 2 AHP Comparison Values

Relative Importance	Value
equal	1
somewhat more important	3
definitely more important	5
much more important	7
extremely more important	9

A pairwise comparison is performed with a quadratic matrix. All diagonal elements are set to 1, and all known relationships (i, j) are filled in. Then, the missing cells (j,i) are each set to $\frac{1}{(i,j)}$. The result is a set of *relative importance tables*. A possible relative importance table for the first level of our example from Fig. 3 is shown in Table 3. For example, "Technical" is judged to be "somewhat more important" than "Currentness", thus cells $(1,2) = 3$ and $(2,1) = \frac{1}{3}$.

Table 3 Relative Importance Table for Technical, Currentness, and Context

	Technical	Currentness	Context
Technical	1	3	1/5
Currentness	1/3	1	1/7
Context	5	7	1

In a simplified version of the computation of the maximum eigenvalue and eigenvectors, the *priority vector* is obtained (for details, see Saaty 1980). Priority vectors are computed for all criteria and for all alternatives considered for each criterion. Afterwards, the intermediate priority vectors for each criterion are merged with the help of the importance vectors of all criteria, to obtain, for each alternative, a score, which is used to build the ranking.

In an 'ideal world', every user would specify her or his individual preferences on attributes. However, in more realistic settings, a system may at best have such preferences from some users and will have to serve unknown/anonymous users on this basis. Thus, an aggregation over weights becomes necessary. Since the users who expressed preferences as well as those who did not will usually be heterogeneous groups, this can lead to non-appropriate rankings. An example is the aggregation of extreme values towards a mean value, which will please no-one. To solve this conflict, a clustering algorithm can be implemented that resolves heterogeneity.

4 Making Randomness Transparent

As we have argued in Section 2.9, user rankings most probably have an element of randomness. As several studies have shown, users however tend to have a strong belief that system rankings (and therefore the "truth"?!) are deterministic, cf. Section 2.8. As a possible remedy, Pan, Hembrooke, Joachims, Lorigo, Gay, and Granka (2007) suggested in their outlook that "... [a] certain degree of randomness in the ranking of returned results ... leads to improved search". In this section, we will investigate the possibilities of introducing randomness into rankings in a more principled way.

The first important observation is that this would have to be communicated – randomness would need to be transparent. Otherwise, search engines will risk to appear "buggy", and some user tasks that employ search engines may become unnecessarily difficult. As an example, consider the exchange of the guideline "Google for XYZ and take the second result". Rankings that present themselves in such a way might also help people remember (more strongly) that context and framing (see Sections 2.2 and 2.5) are ubiquitous and influence rankings.

The randomness of a system's ranking represents the uncertainty of user preferences, which are a result of hidden processes as well as of inconsistencies of the partial ordering. The randomness should therefore be adequately designed, i.e. the parameterization of the underlying distribution function has to be carefully considered. The introduction of a random process should be done as an element of the function Φ as:

$$\Phi^*(\alpha) = \Phi(\alpha) + u, \tag{5}$$

where $u \sim Dist(\cdot)$ is a random variable distributed according to function *Dist*. The implementation of randomness in Φ^* should be small so the ranking of the system S is not completely random, but at local points the ranking might differ for different ranking building processes.

Randomness poses further challenges when users want to share rankings with other people, because the ranking is treated as only one possible ordering. A disadvantage of using randomness is that the user does not re-obtain the ranking when issuing the (identical) query again. However, this is only a disadvantage at a first glance: when pseudo-random numbers are used, their generator uses a seed, which determines the "randomness". If this seed is re-used for another ranking building process, the same ranking will be reproduced. This will enable users to re-use their own searches with identical search results, and it will allow them to share these results by sending the user-specific parameters of Φ including the seed. Obviously, the information thus shared might contain highly privacy-sensitive data: personal preferences together with a seed which, if it is persistent, might be used to uniquely identify the user. Storing and exchanging such information would thus pose new challenges for security and privacy mechanisms.

At present (June 2009), some randomness appears to be introduced into result lists by Google: The ranking varies by browser, browser and search-engine preferences (`http://www.google.com/preferences`), and possibly other factors. A replication of a result list is possible by saving the URL querystring, but users have no further control over this process.

5 Conclusion and Outlook

To evaluate the quality of a ranking such as done in search engines or recommender systems, one must answer the question whether the ranking is the "right ranking"for the given person, in the given circumstances. In this paper, we proposed a very simple general conceptualization of the ranking-evaluation task: the comparison between the ranking generated by a computational system such as a search engine, and the "true ranking inside the user's head". The article then proceeded to describe eight challenges to this simple model. We concluded that they all call for approaches to dealing with multidimensional and often only partial preference orders – both on the part of the users and on the part of the system, and that randomness is probably a characterizing feature of user rankings, and could be a beneficial feature of system rankings. We then proposed that a closer look at existing work in multicriteria decision analysis and the introduction of randomness into system rankings could address the challenges, thus leading to better ranking systems.

As an outlook, we would like to emphasize an additional challenge: several stakeholders. When the interests and preferences of several stakeholders need to be taken into account, the challenges described above also occur. However, recommendation to groups differs from recommendation to individuals (Jameson and Smyth 2007). One important difference is that group members may want to examine one another's preferences. Therefore, the information about preferences has to be acquired. Another task is the determination of the suitability of items for the group and the generation of recommendations from this. Furthermore, the suitability to different members of the group may be different, and recommendations have to be presented in a way that accommodates such differences. Additionally, for a final decision, negotiation may be necessary or required – so the system must help users arrive at a consensus about which recommendations (if any) are acceptable.

Approaches for solving these subtasks are many-faceted and depend on the intention of the system (or the developer of the system). We will only give some examples of preference aggregation for the generation of multi-user rankings. The group ranking may be derived as the minimum, maximum, or average of the rankings of individual group members. Alternatively, a group preference model may be created, for example as a content profile for the group defined as a linear combination of vector-space representations of Web pages that the members like. Finally, sometimes it may be most adequate to issue a recommendation for the group that is not based on the aggregation of preferences of the group's members. An example is a heuristic like "Walt Disney films are good for families" – regardless of whether any person in the family is a particular fan of such movies.

All of these aggregation methods reflect certain strategies and have associated advantages and disadvantages. For example, using the minimum of members' rankings leads to recommending the item that will cause the 'least suffering', while averaging of extreme values may produce a solution that everybody resents as 'a lukewarm compromise'. Many issues in recommendations for groups still need to be resolved. The increasing enrichment of the traditional Web (in which a search engine interacts with an individual user and tailors rankings to her) by the Social Web (in which applications interact with various collectives) offers many opportunities for such research and application building.

References

[Altman and Tennenholtz 2008] Altman, A., Tennenholtz, M.: Axiomatic foundations for ranking systems. Journal of Artificial Intelligence Research 31, 473–495 (2008), `http://www.jair.org/media/2306/live-2306-3748-jair.pdf` (retrieved 2009-06-15)

[Bamberg and Coenenberg 2002] Bamberg, G., Coenenberg, A.G.: Betriebswirtschaftliche Entscheidungslehre, 11th edn. Vahlen, Munich (2002)

[Berendt 2009] Berendt, B.: Ranking – use and usability. To appear in Bulletin of the Belgian Mathematical Society – Simon Stevin (2009)

[Berendt et al. 2005] Berendt, B., Günther, O., Spiekermann, S.: Privacy in e-commerce: Stated preferences vs. actual behavior. Communications of the ACM 48(4), 101–106 (2005)

[Berendt and Kralisch 2009] Berendt, B., Kralisch, A.: A user-centric approach to identifying best deployment strategies for language tools: The impact of content and access language on web user behaviour and attitudes. Journal of Information Retrieval 12(3), 380–399 (2009)

[Berendt and Trümper 2009] Berendt, B., Trümper, D.: Semantics-based analysis and navigation of heterogeneous text corpora: The porpoise news and blogs engine. In: Ting, I.-H., Wu, H.-J. (eds.) Web Mining Applications in E-commerce and E-services, pp. 45–64. Springer, Berlin (2009)

[Chakrabarti 2003] Chakrabarti, S.: Mining the Web. Morgan Kaufmann, San Francisco (2003)

[Dix et al. 1998] Dix, A., Finlay, J., Abowd, G., Beale, R.: Human Computer Interaction. Prentice Hall Europe, Englewood Cliffs (1998)

[d'Ocagne 1885] d'Ocagne, M.: Coordonnées Parallèles et Axiales: Méthode de transformation géométrique et procédé nouveau de calcul graphique déduits de la considération des coordonnées parallèlles. Gauthier-Villars, Paris (1885)

[Eyetools 2008] Eyetools, Eyetools research and reports: Eyetools, enquiro, and did-it uncover search's golden triangle (2008), `http://www.eyetools.com/inpage/research_google_eyetracking_heatmap.htm` (retrieved 2009-06-15)

[Figueira et al. 2005] Figueira, J., Greco, S., Erhgott, M. (eds.): Multiple Criteria Decision Analysis: State of the art surveys. Springer Science and Business Media, Boston (2005)

[Glänzel 2008] Glänzel, W.: Seven myths in bibliometrics about facts and fiction in quantitative science studies. Collnet Journal of Scientometrics and Information Management 2(1), 9–17 (2008), `http://www.collnet.de/Berlin-2008/GlanzelWIS2008smb.pdf` (retrieved 2009-06-15)

[Hearst 2006] Hearst, M.A.: Design recommendations for hierarchical faceted search interfaces. In: SIGIR 2006 Faceted Search Workshop (2006), `http://flamenco.berkeley.edu/papers/faceted-workshop06.pdf` (retrieved 2009-06-15)

[Herlocker et al. 2004] Herlocker, J.L., Konstan, J.A., Terveen, L.G., Riedl, J.T.: Evaluating collaborative filtering recommender systems. ACM Trans. Inf. Syst. 22(1), 5–53 (2004)

[Hu et al. 2007] Hu, J., Zeng, H.-J., Li, H., Niu, C., Chen, Z.: Demographic prediction based on user's browsing behavior. In: WWW 2007: Proceedings of the 16th international conference on World Wide Web, pp. 151–160. ACM, New York (2007)

[Inselberg 1985] Inselberg, A.: The plane with parallel coordinates. Visual Computer 1(4), 69–91 (1985)

[International Organization for Standardization 2007] International Organization for Standardization, ISO 9241-400:2007. ergonomics of human–system interaction – part 400: Principles and requirements for physical input devices (2007), `http://www.iso.org/iso/iso_catalogue/catalogue_tc/catalogue_detail.htm?csnumber=38896` (retrieved 2009-06-15)

[International Organization for Standardization 2008] International Organization for Standardization, ISO 9241-151:2008. ergonomics of human-system interaction – part 151: Guidance on world wide web user interfaces (2008), `http://www.iso.org/iso/iso_catalogue/catalogue_tc/catalogue_detail.htm?csnumber=37031` (retrieved 2009-06-15)

[Jameson 2003] Jameson, A.: Adaptive interfaces and agents. In: Jacko, J.A., Sears, A. (eds.) Human-Computer Interaction Handbook, pp. 305–330. Erlbaum, Mahwah (2003), `http://dfki.de/~jameson/abs/Jameson03Handbook.html` (retrieved 2009-06-15)

[Jameson and Smyth 2007] Jameson, A., Smyth, B.: Recommendation to groups. In: Brusilovsky, P., Kobsa, A., Nejdl, W. (eds.) Adaptive Web 2007. LNCS, vol. 4321, pp. 596–627. Springer, Heidelberg (2007)

[Jansen et al. 2007] Jansen, B.J., Zhang, M., Zhang, Y.: Brand awareness and the evaluation of search results. In: WWW 2007: Proceedings of the 16th international conference on World Wide Web, pp. 1139–1140. ACM, New York (2007)

[Kaplan and Norton 1996] Kaplan, R.S., Norton, D.P.: The Balanced Scorecard. Translating Strategy Into Action. Harvard Business School Press (1996)

[Kaplan and Norton 2004] Kaplan, R.S., Norton, D.P.: Strategy Maps: Converting Intangible Assets Into Tangible Outcomes. Harvard Business School Press (2004)

[Keeney and Raiffa 1976] Keeney, R., Raiffa, H.: Decisions with Multiple Objectives; Preferences and Value Tradeoffs. John Wiley & Sons, Chichester (1976)

[Kobsa and Teltzrow 2005] Kobsa, A., Teltzrow, M.: Impacts of contextualized communication of privacy practices and personalization benefits on purchase behavior and perceived quality of recommendation. In: Beyond Personalization 2005: A Workshop on the Next Stage of Recommender Systems Research (IUI 2005), San Diego, CA, pp. 48–53 (2005)

[Kralisch and Berendt 2005] Kralisch, A., Berendt, B.: Language-sensitive search behaviour and the role of domain knowledge. The New Review of Hypermedia and Multimedia 11(2), 221–246 (2005)

[Kralisch and Köppen 2005] Kralisch, A., Köppen, V.: The impact of language on website use and user satisfaction: Project description. In: Proceedings of the 13th European Conference on Information Systems, Information Systems in a Rapidly Changing Economy, ECIS 2005 (2005)

[Langville and Meyer 2006] Langville, A.N., Meyer, C.D.: Google's PageRank and Beyond: The Science of Search Engine Rankings. Princeton University Press, Princeton (2006)

[Lewandowski and Höchstötter 2007] Lewandowski, D., Höchstötter, N.: Web searching: A quality measurement perspective. In: Spink, A., Zimmer, M. (eds.) Web Searching: Interdisciplinary Perspectives. Springer, Dordrecht (2007)

[Liu and Mihalcea 2007] Liu, H., Mihalcea, R.: Of men, women, and computers: Data-driven gender modeling for improved user interfaces. In: Proceedings of the International Conference on Weblogs Social Media (ICWSM), pp. 121–128 (2007)

[Marchant 2009] Marchant, T.: An axiomatic characterization of the ranking based on the h-index and some other bibliometric rankings of authors. Scientometrics (2009), http://www.springerlink.com/content/e71h95u774701j1k (retrieved 2009-06-15)

[Mozilla Labs 2008] Mozilla Labs, Introducing geode (2008), http://labs.mozilla.com/2008/10/introducing-geode/ (retrieved 2009-06-15)

[Murakami et al. 2008] Murakami, T., Mori, K., Orihara, R.: Metrics for evaluating the serendipity of recommendation lists. In: Satoh, K., Inokuchi, A., Nagao, K., Kawamura, T. (eds.) JSAI 2007. LNCS (LNAI), vol. 4914, pp. 40–46. Springer, Heidelberg (2008)

[Nielsen 2006] Nielsen, J.: Eyetracking research (2006), http://www.useit.com/eyetracking (retrieved 2009-06-15)

[Pan et al. 2007] Pan, B., Hembrooke, H., Joachims, T., Lorigo, L., Gay, G., Granka, L.: In: google we trust: Users' decisions on rank, position, and relevance. Journal of Computer-Mediated Communication 12(3), 801–823 (2007)

[Radlinski and Joachims 2005] Radlinski, F., Joachims, T.: Query chains: learning to rank from implicit feedback. In: Grossman, R., Bayardo, R.J., Bennett, K.P. (eds.) Proceedings of the Eleventh ACM SIGKDD International Conference on Knowledge Discovery and Data Mining, pp. 239–248. ACM, New York (2005)

[Robertson and Zaragoza 2007] Robertson, S., Zaragoza, H.: On rank-based effectiveness measures and optimization. Inf. Retr. 10(3), 321–339 (2007)

[Rousseau 2008] Rousseau, R.: Woeginger's axiomatisation of the h-index and its relation to the g-index, the $h^{(2)}$-index and the R^2-index. Journal of Informetrics 2(4), 335–340 (2008)

[Ruthven and Lalmas 2003] Ruthven, I., Lalmas, M.: A survey on the use of relevance feedback for information access systems. Knowl. Eng. Rev. 18(2), 95–145 (2003)

[Saaty 1980] Saaty, T.: The Analytic Hierarchy Process for Decisions in a Complex World. McGraw-Hill, New York (1980)

[Schneeweiß 1991] Schneeweiß, C.: Planung 1, Systemanalytische und entscheidungstheoretische Grundlagen. Springer, Berlin (1991)

[Sørensen et al. 2006] Sørensen, P., Lerche, D., Thomsen, M.: Developing decision support based on field data and partial order theory, pp. 259–283. Springer, Berlin (2006)

[Tullis and Albert 2008] Tullis, T., Albert, W.: Measuring the User Experience: Collecting, Analyzing, and Presenting Usability Metrics. Morgan Kaufmann, San Francisco (2008)

[Tversky and Kahneman 1981] Tversky, A., Kahneman, D.: The framing of decisions and the psychology of choice. Science 211, 453–458 (1981)

[Varian 2007] Varian, H.R.: Intermediate Microeconomics: A Modern Approach, 7th edn. W W Norton & Co. (2007)

[Vaughan 2004] Vaughan, L.: New measurements for search engine evaluation proposed and tested. Information Processing & Management 40(4), 677–691 (2004)

[Zhang and Hurley 2009] Zhang, M., Hurley, N.: Statistical modeling of diversity in top-n recommender systems. In: To appear in Proc. of the ACM Web Intelligence Conference WI 2009 (2009)

Data Mining on Folksonomies

Andreas Hotho

Abstract. Social resource sharing systems are central elements of the Web 2.0 and use all the same kind of lightweight knowledge representation, called *folksonomy*. As these systems are easy to use, they attract huge masses of users. Data Mining provides methods to analyze data and to learn models which can be used to support users. The application and adaptation of known data mining algorithms to folksonomies with the goal to support the users of such systems and to extract valuable information with a special focus on the Semantic Web is the main target of this paper.

In this work we give a short introduction into folksonomies with a focus on our own system BibSonomy. Based on the analysis we made on a large folksonomy dataset, we present the application of data mining algorithms on three different tasks, namely spam detection, ranking and recommendation. To bridge the gap between folksonomies and the Semantic Web, we apply association rule mining to extract relations and present a deeper analysis of statistical measures which can be used to extract tag relations. This approach is complemented by presenting two approaches to extract conceptualizations from folksonomies.

1 Introduction

Complementing the Semantic Web effort, a new breed of so-called "Web 2.0" applications recently emerged on the Web. These include user-centric

Andreas Hotho
Knowledge & Data Engineering Group,
University of Kassel,
34121 Kassel, Germany
e-mail: hotho@cs.uni-kassel.de

G. Armano et al. (Eds.): Intelligent Information Access, SCI 301, pp. 57–82.
springerlink.com

publishing and knowledge management platforms like wikis, blogs, and social resource sharing tools. In this paper, we focus on resource sharing systems, which all use the same kind of lightweight knowledge representation, called *folksonomy*.[1]

Social resource sharing systems are web-based systems that allow users to upload all kinds of resources, and to label them with arbitrary words, so-called *tags*. The systems can be distinguished according to what kind of resources are supported. Flickr[2], for instance, allows the sharing of photos, del.icio.us[3] the sharing of bookmarks, CiteULike[4] and Connotea[5] the sharing of bibliographic references, and 43Things[6] even the sharing of goals in private life. Our own system, *BibSonomy*,[7] allows sharing bookmarks and BibTeX entries simultaneously.

According to Fayyad et. al. *Data Mining* is "the nontrivial process of identifying valid, previously unknown, and potentially useful patterns" [12] in a potentially very huge amount of data. Web Mining is the application of data mining techniques on three areas: the content, the structure and the usage of resources in the web [33]. Although Web 2.0 systems – as the name suggests – are still web applications and the analysis of such systems could be subsumed under the term web mining, new challenges for data mining emerge, as new structures and new data can be found in such systems. Therefore, we call the analysis of folksonomies: *Folksonomy Mining*. One example is that structure mining is applied on folksonomies and not – as it is known from web mining – on the web graph as a whole. Given the high number of publications in the short lifetime of folksonomy systems, researchers seem to be very interested in folksonomies and the information and knowledge which can be extracted from them. This can be explained by the tremendous amount of information collected from a very large user basis in a distributed fashion in such systems.

The application of mining techniques on folksonomies bears a large potential. Further, it is in line with the general idea of *Semantic Web Mining* [47]. Two aspects are of central interest: On the one hand, folksonomies form a rich source of data which can be used as a source for full-blown ontologies. This process is known as ontology learning and often utilizes data mining techniques. On the other hand, mining the Semantic Web is a second important application of mining techniques in this area. As folksonomies are considered as weak knowledge representation, analyzing their data can be seen as an implementation of Semantic Web Mining. The goal of this work is therefore to bridge the gap between folksonomies and the Semantic Web and to start to solve this problem with research contributions from various

[1] `http://www.vanderwal.net/folksonomy.html`
[2] `http://www.flickr.com/`
[3] `http://delicious.com`
[4] `http://www.citeulike.org`
[5] `http://www.connotea.org`
[6] `http://www.43things.com`
[7] `http://www.bibsonomy.org`

sides. More precisely, to reach this goal, a better understanding of the hidden and emergent semantics in folksonomies is necessary, as well as methods to extract the hidden information. Data Mining techniques provide methods for solving these issues.

This paper gives an overview of the previously published articles [24, 7, 34, 25, 31, 41, 5, 30] and shows connection between them. There are two lines of research: We analyzed the data we collected in order to get a better understanding of its structure (cf. [7]), and developed algorithms to support the users of folksonomy systems (cf. [25, 34, 31]). Second, we developed our own system BibSonomy as a platform for research experiments (cf. [24]). As we own the system, we have full access on e.g. all data, the user interface and so on. This puts us in the situation which researchers usually do not have: We can perform research experiments to test our new methods and push our research results into BibSonomy to show, to evaluate, and to demonstrate the advantages of our methods. One example are the online recommender experiments we are doing for this year's ECML PKDD discovery challenge.[8] Further, we have implemented many of our research results from the last years into the system. One of the first results having found its way into BibSonomy was a lightweight recommender (cf. Sec. 3.3) followed by the FolkRank ranking (cf. Sec. 3.2).

Related Work

Folksonomies and especially data mining on folksonomies are a relatively young research area. Meanwhile, work for specific areas starts to show up. To discuss the related work for all methods mentioned in this paper is beyond the scope of it. More detailed surveys can be found in the respective papers. To start with folksonomies and to learn more about their strengths and weaknesses one may look into [19, 36, 37]. One of the first works defining a model of semantic-social networks for extracting lightweight ontologies from del.icio.us was [38]. Recently, work on more specialized topics such as structure mining on folksonomies – e. g. to visualize trends [11] and our work on patterns [41] in users' tagging behavior – as well as ranking of folksonomy contents [25], analyzing the semiotic dynamics of the tagging vocabulary [6], or Halpin's analysis of the dynamics and semantics [18] have been presented.

Structure of the Paper

The rest of the paper is structured as follows: After a short introduction into the topic of social bookmarking and folksonomies we present a formal model and first properties we found in the graph formed by folksonomies. In Section 3 we present the three applications spam detection, a ranking method for folksonomies and a tag recommender method. Section 4 goes

[8] `http://www.kde.cs.uni-kassel.de/ws/dc09/`

one step towards more semantics in folksonomies and presents approaches to bridge the gap between folksonomies and the Semantic Web.

2 Basics of Folksonomies

In this section we will review the basic principles of folksonomies. We will start with an introduction into folksonomies followed by the description of our own system BibSonomy. A formal definition and first insights into the properties of such systems are the next part of this section. The presented ideas are the basis for the following steps where we first use these insights to provide valuable services like a better ranking or tag recommendation before we investigate the extraction of semantics out of folksonomies.

2.1 Social Bookmarking Systems

First bookmarking systems were developed at the end of the 90s but without a nice and fast user interface and often with a very weak business model (cf. [22]). Therefore, it was virtually impossible to attract a large number of users – which is necessary to make such systems attractive. One of the first systems which could reach a broad user basis was del.icio.us.[9] It was started in 2003 by Joshua Schachter and is today the best-known social bookmarking system for websites in the world. After he released a first version in 2003, he followed the advice of his users to make it more attractive. In 2004 the system reached a critical mass and the number of users increased dramatically. At the end of 2005 he sold the system to Yahoo. It is still running and the number of users is estimated with more than five million.[10] Similar services followed and provided a comparable service. Some of them focus on different content types, e.g. images, music, videos or places. Others provide an added value in form of additional functionality, e.g. by caching the seen webpage or presenting improved tag clouds for easier browsing. The core structure of all these systems is very similar and is known under the name *folksonomy*.

For research purposes we collected data from del.icio.us. The first time we crawled it was in 2005, where we collected the complete user pages for more than 75000 users. At that time we were able to gather an almost complete snapshot (mostly without spam). The second time we crawled del.icio.us in 2006 and collected data of more than 600000 users. Within our research we used and use these two datasets for our analyses and scientific experiments. Details can be found in the cited works.

In the next section, we will describe our own social bookmark and publication sharing system BibSonomy before we focus on the core structure of social bookmarking systems.

---

[9] `http://delicious.com/`

[10] `http://blog.delicious.com/blog/2008/11/delicious-is-5.html`

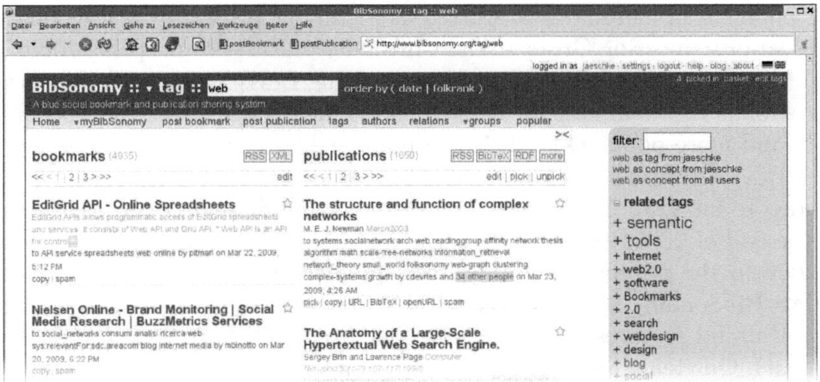

Fig. 1 BibSonomy displays bookmarks and BibTeX based bibliographic references simultaneously

2.2 BibSonomy: A Social Bookmark and Publication Sharing System

Resource sharing systems like BibSonomy provide an easy way to organize and manage different kinds of resources. What is considered as a resource depends on the type of system. For instance, in del.icio.us, the resources are URLs, in flickr, the resources are pictures, and in BibSonomy they are either URLs or publication entries. As described in the previous section, in their core, these systems are all very similar. Once a user is logged in, he can add a resource to the system, and assign arbitrary tags to it. The collection of all his assignments is his *personomy*, the collection of all personomies constitutes the *folksonomy*. As in other systems, the user can explore his personomy, as well as the personomies of the other users, in all dimensions: for a given user one can see all resources he has uploaded, together with the tags he has assigned to them; when clicking on a resource one sees which other users have uploaded this resource and how they tagged it; and when clicking on a tag one sees who assigned it to which resources (see Figure 1). The systems allows for additional functionality. For instance, one can copy a resource from another user, and label it with one's own tags. Overall, these systems provide a very intuitive navigation through the data.

BibSonomy[11] is one of the social resource sharing tools that have acquired large numbers of users within the last years. The reason for their immediate success is the fact that no specific skills are needed for participating, and that these tools yield immediate benefit for the individual user (e.g. organizing ones bookmarks in a browser-independent, persistent fashion) without too much overhead. Additionally, BibSonomy allows to share both bookmarks

[11] http://www.bibsonomy.org/

and publication metadata. It started as a student project at the Knowledge and Data Engineering Group of the University of Kassel[12] in spring 2005. The goal was to implement a system for organizing BIBTEX entries (cf. [39]) in a way similar to bookmarks in del.icio.us – which was at that time becoming more and more popular. BIBTEX is a popular literature management system for LATEX, which many researchers use for writing scientific papers. We soon decided to integrate bookmarks as a second type of resource into the system. At the end of 2005, we announced BibSonomy first to some colleagues, later in 2006 to the public. Since then, the number of users has grown steadily. Today, BibSonomy has more than 190000 registered users. We implemented several useful features and redesigned the architecture to ease future developments. Our team and other research groups use BibSonomy or its data for research, and we have implemented our research results into the system, e.g. the FolkRank algorithm and tag recommendation methods – both for the benefit of the users and to directly measure the performance of our methods. A more detailed description of BibSonomy can be found in [24]. In the following subsections, we will give pointers to the most interesting parts of the system.

2.2.1 User Interface

A typical list of posts is depicted in Figure 1 which shows bookmark and publication posts containing the tag *web*. The page is divided into four parts: the header (showing information such as the current page and path, navigation links and a search box), two lists of posts – one for bookmarks and one for publications – each sorted by date in descending order, and a list of tags related to the posts. This scheme holds for all pages that are showing posts; it allows for navigation in all dimensions of the folksonomy. The posts in the lists are sorted by date in descending order, while the tags can be sorted lexicographically or by frequency of usage, depending on the user's choice.

Beside this kind of pages, systems like BibSonomy typically contain summary pages representing the content in form of a cloud. The page with the global tag cloud summarizes in a clear way the content of the system by the used tags. A similar functionality is offered by the author and relation pages which are special pages for BibSonomy. Note that on selected pages posts can be ordered by relevance as calculated by the FolkRank algorithm (cf. Sec. 3.2).

2.2.2 Architecture

The basic building blocks of BibSonomy are an Apache Tomcat[13] servlet container using Java Server Pages[14] and Java Servlet[15] technology and a

[12] `http://www.kde.cs.uni-kassel.de/`
[13] `http://tomcat.apache.org/`
[14] `http://java.sun.com/products/jsp`
[15] `http://java.sun.com/products/servlets`

MySQL[16] database as backend. The project uses the Model View Controller (MVC) programming paradigm to separate the logical handling of data from the presentation. This enables us to produce output in various formats (see Section 2.2.3), since adding a new output format is accomplished by simply implementing a JSP as a view of the model.

The central database schema of BibSonomy is based on four tables: one for bookmark posts, one for publication posts, one for tag assignments (*tas*) and one for *relations* between tags. Two further tables store information regarding *users* and *groups*.

The post tables are connected with the tas table by the key *post_id*. The schema is not normalized – on the contrary we have added a high amount of redundancy to speed up queries. For example, besides storing group, user name and date in the posts table, we also store it in the tas table to minimize the number of rows touched when selecting rows for the various views. Furthermore, several other tables hold counters (i. e., how many people share one resource, how often a tag is used, etc.). Finally, a large set of indexes (12 in the tas table alone) builds the basis for a fast answering of queries.

Overall, we spent a large amount of work on investigating and optimizing SQL queries and table schemas and tested both with folksonomy data of up to 8000000 posts. At the moment, we need no special caching or physical distribution of the database to get reasonable response times.

2.2.3 Features

The most simplistic but also most laborious way to add posts to BibSonomy is by entering their metadata manually into form fields. To lower the effort to get data into BibSonomy, it supports various ways to import resources from files and web pages (e.g. BibTeX or Endnote[17]) or by so called "scrapers"[18] which allow to automatically extract publication metadata from digital libraries like SpringerLink.[19] Nevertheless – forms are still used to edit posts.

Exporting publication references in BibTeX format is accomplished by preceding the path of a URL showing publication posts with the string `/bib` – this returns all publications shown on the respective page in BibTeX format. For example the page `http://www.bibsonomy.org/bib/search/text+clustering` returns a BibTeX file containing all literature references which contain the words "text" and "clustering" in their fulltext.

More general, every page which shows posts can be represented in several different ways by preceding the path of the URL with a specific string to specify the export format, e.g. `/xml` for bookmarks in XML format or `/publ` for publications in a simple HTML format suited for the integration into a

[16] `http://www.mysql.com/`
[17] `http://www.endnote.com/`
[18] `http://scraper.bibsonomy.org/`
[19] `http://www.springerlink.de/`

web page (for an integration example see http://www.kde.cs.uni-kassel.de/pub). For an overview of the available export formats for publications, one can use the /export path extension which is also linked on all web pages showing publication posts. The export feature allows to generate publication lists for external websites, e.g. for personal and institute webpages or for project pages.

Experience has shown that an Application Programming Interface (API) is crucial for a Web 2.0 system to gain success. Hence we have implemented a lightweight REST API[20] which can be used and accessed also by less experienced programmers. We use the API for the integration of JabRef.[21] The catalogue of the library of the university of Cologne uses the API to access tagging information for its books.

There are several other valuable features like the publication basket, the duplicate detection mechanism, a tag editor, the mirror of the famous DBLP computer science library,[22] or the integration with other systems. A description of those features can be found in [24]. In our blog,[23] we report regularly on new developments. For research purposes, we release a complete snapshot of BibSonomy's public data on a regular basis.[24]

2.3 Folksonomies

As described in Sec. 2.1, folksonomies are the core structure of social bookmarking systems. The word "folksonomy" is a blend of the words "taxonomy" and "folk", and stands for conceptual structures created by the people.[25] The way an folksonomy is emerging is the same in all these systems and can be described as follows: There is a user who is interested in a certain resource. A folksonomy system provides a way to store this resource and to annotate it. Typically, the annotation process is as simple as possible and driven by keywords called *tags*.[26] The tags can serve several purposes [16], e.g. they describe the content of a resource or the reasons why the resource was saved. The central elements of a folksonomy are depicted in Figure 2. The center is formed by a post which connects a user with tags and a resource. Different users can use different tags to describe the same resource and resources are typically tagged by several users. The emergent structure of users, tags and resources is called *folksonomy*.

---

20 http://www.bibsonomy.org/help/doc/api.html
21 http://bibsonomy.blogspot.com/2009/02/feature-of-week-bibsonomy-plugin-for.html
22 http://www.informatik.uni-trier.de/~ley/db/
23 http://bibsonomy.blogspot.com/
24 http://www.bibsonomy.org/help/faq/600_benchmark
25 http://www.vanderwal.net/folksonomy.html
26 A more exotic example is the use of geographic coordinates as tags to describe where a photo was taken. In principle this is the same annotation process.

Such systems provide direct benefits which makes them attractive for their users. The process of tagging is very simple and it is very easy for every user to access his own resources. The web based storage of e.g. bookmarks allows to access them at any time from all over the world. Bookmarks are no longer tied to a single computer. Due to the nature of such systems, browsing within the bookmark collection leads to the serendipitous discovery of unknown resources. This property of folksonomies is mostly unexpected by its users, but it makes systems with a folksonomy core so fascinating. Despite the uncoordinated tag assignment of different users in such systems, the emergence of semantics can be observed (details in Sec. 2.5).

Another advantage of folksonomies is the human-contributed annotation which can be seen as a lightweight knowledge representation. Most of the tags describe the content of the annotated resource and as they are assigned by humans which are able to grasp the content, the resulting description is better than automatic solutions used in search engines or categorizations systems. However, the broad range of human contribution is also a major disadvantage of folksonomies. To make the usage of bookmarking system simple, it is allowed to use any arbitrary tag in a totally uncontrolled way (cf. [48]). This results in difficulties, as tags tend to suffer from typical language problems like synonyms, polysemy and singular vs. plural forms. The usage of tags can be driven by a very personal preferences which confuses others and does not contribute to common semantics.

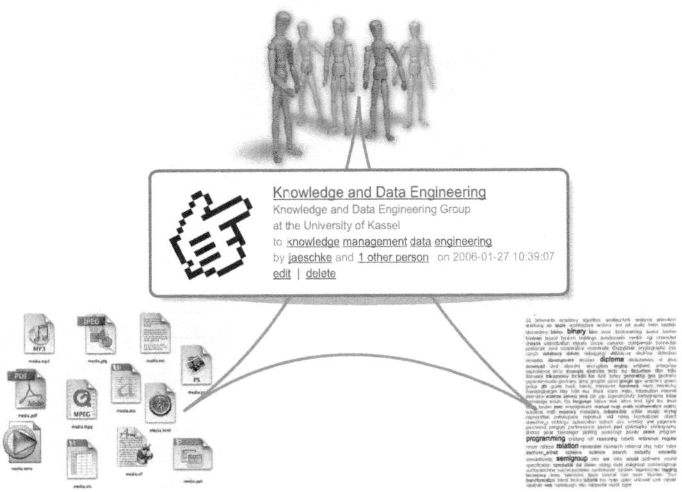

Fig. 2 Visualization of an example bookmark post of a tagging system

2.4 A Formal Model for Folksonomies

A folksonomy describes the users, resources, and tags, and the user-based assignment of tags to resources. We present here a formal definition of folksonomies (cf. [25]), which is also underlying our BibSonomy system.

Definition 1. A *folksonomy* is a tuple $\mathbb{F} := (U, T, R, Y, \prec)$ where

- U, T, and R are finite sets, whose elements are called *users*, *tags* and *resources*, resp.,
- Y is a ternary relation between them, i. e., $Y \subseteq U \times T \times R$, whose elements are called tag assignments (*tas* for short), and
- $\prec$ is a user-specific subtag/supertag-relation, i. e., $\prec \subseteq U \times T \times T$, called *is-a relation.*

Definition 2. The *personomy* $\mathbb{P}_u$ of a given user $u \in U$ is the restriction of $\mathbb{F}$ to u, i. e., $\mathbb{P}_u := (T_u, R_u, I_u, \prec_u)$ with $I_u := \{(t, r) \in T \times R \mid (u, t, r) \in Y\}$, $T_u := \pi_1(I_u)$, $R_u := \pi_2(I_u)$, and $\prec_u := \{(t_1, t_2) \in T \times T \mid (u, t_1, t_2) \in \prec\}$, where π_i denotes the projection on the ith dimension.

Users are typically described by their user ID, and tags may be arbitrary strings. What is considered as a resource depends on the type of system, e.g. in BibSonomy they are either URLs or publication entries.

Definition 3. For convenience we also define the set P of all *posts* as

$$P := \{(u, S, r) \mid u \in U, r \in R, S = \text{tags}(u, r), S \neq \emptyset\}$$

where, for all $u \in U$ and $r \in R$, $\text{tags}(u, r) := \{t \in T \mid (u, t, r) \in Y\}$ denotes all tags the user u assigned to the resource r.

If we disregard the is-a relation, we can simply note a folksonomy as a quadruple $\mathbb{F} := (U, T, R, Y)$. This structure is known in Formal Concept Analysis [50, 15] as a *triadic context* [35, 46]. An equivalent view on this structure is that of a tripartite (undirected) hypergraph $G = (V, E)$, where $V = U \dot\cup T \dot\cup R$ is the set of nodes, and $E = \{\{u, t, r\} \mid (u, t, r) \in Y\}$ is the set of hyperedges.

In a typical folksonomy system, every tag assignment is connected with several other properties like date, group or resource type. For sake of simplicity, we disregard these properties for the rest of the work, unless stated otherwise.

2.5 Network Properties of Folksonomies

The new data of folksonomy systems provides a rich resource for data analysis, information retrieval, and knowledge discovery applications. We made a first step towards this end in [7]. The goal is to gain better insights into these systems by analyzing the main network characteristics on two example systems.

To this extent, we investigate the growing network structure of folksonomies over time from different viewpoints, using two datasets from del.icio.us and BibSonomy as examples. First, we investigate the network structure of folksonomies much on the same line as the developments in the research area of complex networks. To that end, we adapt classical network measures like characteristic path length and clustering coefficient for so-called "small world networks" which have been used on a wide variety of graphs in recent years, to the particular tripartite structure of folksonomies. We show that folksonomies do indeed exhibit a small world structure, as we observe in both systems very short path length, (in average around 3) in both systems and a very high clustering coefficient compared to random graphs. This helps to explain to some extent the successful serendipitous browsing of users in such systems. The small world property implies for the users that only few clicks are needed to end up in a new and hopefully interesting topic within a folksonomy. On the other hand, the high clustering coefficient hints a cluster of resources with a similar topic within the direct neighborhood.

Second, beyond the analysis of the whole hypergraph, we also consider specific projections of it by narrowing the scope and focusing on particular features of the structure. We introduced a weighted network of tags where link strengths are based on the frequencies of the tag-tag co-occurrence, and studied the weight distributions and connectivity correlations among nodes in this network. Our analysis and experiments indicate the existence of the emergence of shared semantics in the folksonomy system, implicitly negotiated by users. We find indicators for both hierarchical and social structures in the network of tag-tag co-occurrence.

Our experiments hint that spam – which becomes an increasing nuisance in social resource sharing systems – systematically shows up in the connectivity correlation properties of the weighted tag-tag co-occurrence network. These activities in data from del.icio.us in its early days indicate the need to develop more advanced methods to fight against such misuse of folksonomy systems. We will present our approach to detect spam in Sec. 3.1. A deeper analysis of the emergent semantics in folksonomies appears promising, and results in this direction are presented in Sec. 4.3. A first application to support the user based on the collaborative intelligence hidden in folksonomies is the tag recommender (cf. Sec. 3.3).

3 Applications

3.1 Spam Detection

Web spam detection is a well known challenge for search engines. Spammers add specific information to their web sites that solely serves the purpose to increase the rank of a page in search results, but not its quality or content. They thereby increase the traffic to their web sites – be it for commercial or political interests or to disrupt the service provided. Ranking algorithms need to detect those pages.

Not only search engines struggle with malicious web content. Social bookmarking systems also have become an attractive place for posting web spam. Spammers (mis)use the popularity and the high ranking of social bookmarking systems in search engines for their purposes. All they need is an account; then they can freely post entries which bookmark the target spam web site. In recent months, different spamming techniques have been developed to frequently show up on popular sites, recent post sites or as highly ranked posts on a search for a specific tag. For instance, spammers register several accounts and publish the same post several times. Besides appearing on the "recent post" page, the bookmark may show up on the "popular page", since "many" users have considered the bookmark. Another technique is to add diverse tags to the bookmark or use popular tags.

In order to retain the original benefits of social bookmarking systems, we developed techniques which prevent spammers from publishing in these systems [34]. The problem can be considered as a binary classification task. Based on different features that describe a user and his posts, a model is built from training data to classify unknown examples (on a post or user level) either as "spam" or "non-spam". As we consider "social" systems in which users interact with each other and one incentive to use the system is to see and to be seen, an exclusion of non-spammers from publishing is a severe error which might prevent the user from further participation. Similar to other spam detection settings, this problem needs to be taken into consideration when classifying users.

The adaptation of classification algorithms to this task consists of two major steps. The first one is to select features for describing the users. The second step is the selection of an appropriate classifier for the problem. In [34], we introduce a set of initial features that can be used for spam classificiation. These features are evaluated with well-known classifiers (SVM, Naive Bayes, J48 and logistic regression) against a simple baseline of representing a user by the usage of tags. Combining all features shows promising results exceeding the AUC and F1 measure of the selected baseline. Considering the different feature groups, co-occurrence features show the best ROC curves.

Our results support the claim of [21], that the problem can be solved with classical machine learning techniques – although not perfectly. The difference to web spam classification are the features applied: on the one hand, more information (e. g., IP address, tags) is given, on the other hand spammers reveal their identity by using a similar vocabulary and similar resources. This is why co-occurrence features tackle the problem very well.

Overall, our contribution represents a first step towards the elimination of spam in social bookmarking systems using machine learning approaches. We implemented a framework within BibSonomy following the results of our analysis. The framework automatically flags in average more than 200 new spammers per day. Besides the practical need to eliminate spam, we intend to use this platform to develop and evaluate further social spam detection mechanisms and to tune the performance of the running machine learning approaches. For this we use also results from last year's ECML PKDD discovery challenge[27] organized by us, where one task was the prediction of spam in social bookmarking systems.

3.2 Ranking in Folksonomies

In the past, folksonomies were able to attract a large number of users who created huge amounts of information. But with the growing number of resources stored within each users personomy, it becomes more and more difficult for the user to find and retrieve the saved resources. A first step to searching folksonomy based systems – complementing the browsing interface usually provided as of today – is to employ standard techniques used in information retrieval or, more recently, in web search engines. Since users are used to web search engines, they likely will accept a similar interface for search in folksonomy-based systems. The research question is how to provide suitable ranking mechanisms, similar to those based on the web graph structure, but now exploiting the structure of folksonomies instead. To this end, we proposed in [25] a new algorithm, called *FolkRank*, that takes into account the folksonomy structure for ranking users, tags and resources in folksonomy based systems. Further, the algorithm can be used for a topic-specific ranking.

The general idea of FolkRank is as follows: Given a set of preferred tags, users, and/or resources, a topic specific ranking provides an ordering of the elements of the folksonomy in descending importance with respect to the preferred elements. To that end, FolkRank is a differential approach of a weight-spreading algorithm which compares the resulting rankings with and without preference vector computed on the folksonomy graph. We implemented the weight-spreading ranking scheme on folksonomies in two steps. First, we transform the hypergraph between the sets of users, tags, and resources into an undirected, weighted, tripartite graph. On this graph, we apply a version of PageRank that takes into account the edge weights.

The original formulation of PageRank [3] reflects the idea that a page is important if there are many pages linking to it, and if those pages are important themselves. We employ a similar motivation for our ranking scheme in folksonomies. The basic notion is that a resource which is tagged with important tags by important users becomes important itself. The same holds, symmetrically, for tags and users, thus we have a tripartite graph in which the vertices are mutually reinforcing each other by spreading their weights. It turned out, however, that running an adapted PageRank as is returned results that were largely dominated by the global structure of the folksonomy, yielding the same top elements such as the tags "web" or "blog" on top no matter what the preferences were. Thus, FolkRank circumvents that problem

[27] `http://www.kde.cs.uni-kassel.de/ws/rsdc08`

using a differential approach. It computes a topic-specific ranking in a folksonomy by computing the winners and losers of the mutual reinforcement of resources when a user preference is given, compared to the baseline without a preference vector. More details can be found in [25].

There, the FolkRank ranking scheme has been used to generate personalized rankings of the items in a folksonomy, and to recommend users, tags and resources. Top folksonomy elements which are retrieved by FolkRank tend to fall into a coherent topic area, e.g. "Semantic Web". This leads naturally to the idea of extracting *communities of interest* from the folksonomy, which are represented by their top tags and the most influential persons and resources. This idea found its way into BibSonomy. There is now an option to rank resources not only by date, but also by FolkRank[28]. The shown page displays not only the ranked resource list but also the top ranked similar tags and the most influential users of this topic. These users form some kind of community. By making such an implicit existing communities explicit, interested users can find other users, also interested in the search topic and in this way community members can more easily get to know each other and learn of others' resources.

When folksonomy-based systems grow larger, user support has to go beyond enhanced retrieval facilities. Therefore, the internal structure has to become better organized. An obvious approach for this are Semantic Web technologies. The key question remains, though, how to exploit its benefits without bothering untrained users with its rigidity. This could be done by utilizing the strength of the semantic technology within a folksonomy system and using data mining methods to bridge the gap between both worlds. One approach going in this direction is presented in Sec. 3.3 as tag recommenders simplify the posting process and in Sec. 4.3 where different kinds of related tags are extracted which can form a basis for a better organization of tags. We believe that this will become a fruitful research area for the Semantic Web community for the next years.

One application of FolkRank is presented in [26]. There, we analyze the emergence of common semantics by exploring trends in the folksonomy. Since the structure of a folksonomy is symmetric with respect to the dimensions "user", "tag", and "resource", we can apply the same approach to study upcoming users, upcoming tags, and upcoming resources over time. With FolkRank, we compute topic-specific rankings on users, tags, and resources. In a second step, we can then compare these rankings for snapshots of the system at different points in time. We can discover both the absolute rankings (who is in the Top Ten?) and winners and losers (who rose/fell most?). We present a technique for analyzing the evolution of topic-specific trends.

Furthermore, there has been a lively discussion in e.g. the delicious-discuss mailing list about the usefulness of the $\prec$ relation in the folksonomy, which is partially realized as *bundles* in del.icio.us. We will investigate first steps to be

[28] For an example, see `http://www.bibsonomy.org/tag/Semantic_Web?order=folkrank`

able to make use of ontology learning techniques to populate this relation in BibSonomy and augment the underlying semantic structure in the folksonomy in Sec. 4.

3.3 Recommending Tags

Recommenders are a common technique to support users in finding new and interesting items, e.g. movies, books, or other products. In folksonomies, recommenders can be used to recommend similar users, interesting resources or help to find the right tags while posting a new resource. We focus on the third task in this section. One example for a resource recommender can be found in [49]. The literature on tag recommendations in folksonomies is still sparse. The existing approaches usually lie in the area of collaborative filtering and information retrieval. Most recently, the ECML PKDD 2008 Discovery Challenge[29] organized by our research group has addressed the problem of tag recommendations in folksonomies [23]. The provided dataset gives a good basis for the research in this area and the upcoming next challenge[30] shows the need for better recommender approaches and increasing interest of researchers in this area.

To support users in the tagging process and to expose different facets of a resource, most of the systems offered some kind of tag recommendations already at an early stage. Del.icio.us, for instance, had a tag recommender in June 2005 at the latest,[31] and also included resource recommendations.[32] However, no algorithmic details were published. We hypothesize that these recommendations basically provide those tags which were most frequently assigned to the resource.

As of today, nobody has empirically shown the benefits of recommenders in such systems. In [31], we evaluate a tag recommender based on Collaborative Filtering, a graph based recommender using our ranking algorithm FolkRank, and several simple approaches based on tag counts. With this research we start a qualitative comparison of different recommender approaches while simultaneously adopting state of the art techniques to work with the underlying triadic graph. The results presented in [31] built upon results presented at ECML PKDD 2007 [32].

The presented results in [31] show that the graph-based approach of FolkRank is able to provide tag recommendations which are significantly better than those of approaches based on tag counts and even better than those of state-of-the-art recommender systems like Collaborative Filtering.

---

[29] `http://www.kde.cs.uni-kassel.de/ws/rsdc08/`

[30] `http://www.kde.cs.uni-kassel.de/ws/dc09/`

[31] `http://www.socio-kybernetics.net/saurierduval/archive/2005_06_01_archive.html`

[32] `http://blog.del.icio.us/blog/2005/08/people_who_like.html`

The tradeoff is that the computation of FolkRank recommendations is cost-intensive so that one might prefer less expensive methods to recommend tags in a social bookmarking system. The *most popular tags ρ–mix* approach proposed by us in [31] has proven to be considered as a solution for this problem. It provides results which can almost reach the quality of FolkRank but which are rather cheap to generate. Especially the possibility to use index structures (which databases of social bookmarking services typically provide anyway) makes this approach a good choice for online recommendations. Finally, despite its simplicity and non-personalized aspect, the *most popular tags* achieved reasonable precision and recall on the small datasets (last.fm and BibSonomy) which indicates its adequacy for the cold start problem of young systems.

One result of the ECML PKDD discovery challenge 2008 was the insight that two recommendation tasks can be distinguished. In [31], we focus on the dense part of the folksonomy. We assume that we have information about both the user and the resource and make use of this information to predict the tags the user will use to describe the resource which was already tagged by other users of the system. Contrary to this, most often not all information is available. This means that either the user or the resource or both are new. In this case, one cannot apply the methods described in [31]. We address this issue in [28] where we utilize the content of the webpage which the user will tag in a content based recommender. The underlying methods are known as text classification approaches. J. Illig evaluates the applicability of these methods in general in [27]. In principle, the application of text classification approaches is possible, but the approaches need to be better adapted to the underlying problem. The high number of classes decreases the performance in terms of runtime behavior and accuracy. An interesting next step is the integration of user information in the recommendation process.

4 Towards More Semantics in Folksonomies

As mentioned in Section 2.2, BibSonomy provides the possibility to store tag relations as a kind of conceptualization. One outcome of Sec. 2.5 is the existence of *emergent semantics* [42, 44] in folksonomies. In this section, we present three approaches which will help to understand and to extract the semantics that are implicitly added by the user and hidden in the folksonomy. We will show ways to make it explicit and available for further use. We start with a short comparison of folksonomies and ontologies.

4.1 Folksonomies and Ontologies

Ontologies are a well-known formalism to represent knowledge in a structured way [43] and are the building block of the "Semantic Web" effort. With their well-defined semantics, ontologies offer benefits for a wide spectrum of applications supported by advanced tools from industry and academics. Nevertheless, there are problems to make use of Semantic Web technology in very large application contexts, especially in the web. The web contains huge masses of data but not in any case the data is available in the structured form needed by the Semantic Web, e.g. as ontologies. The knowledge acquisition bottleneck characterize the phenomena that the transformation process from unstructured to structured information is possible but does not scale to the size of the web. The reason is that a certain expertise is needed to create ontologies and to maintain them. This raises the cost of knowledge acquisition and only few people are contributing. Learning ontologies from text [9] is a first way to simplify the acquisition process by utilizing machine learning approaches and linguistic knowledge.

Folksonomies can be seen as a lightweight knowledge representation. Many unexperienced users contribute small pieces of information – unfortunately only in a weakly structured fashion. There is a large amount of information, but it is unstructured and therefore incompatible with semantically rich representations. Both approaches could benefit from each other: While folksonomies need more structure, ontologies need more contributors. Research in this direction has been stimulated in form of the "Bridging the Gap between Semantic Web and Web 2.0" workshop,[33] where the contributions ranged from the use of human contributed information to simplified Web 2.0-like Semantic Web tools.

The emergent semantics in folksonomies can be extracted by using machine learning algorithms or advanced analysis methods. A first approach is presented in the next section. To be able to develop advanced knowledge extraction methods, a better understanding of the kind of underlying semantics is needed. We presented the summary of a first analysis in Sec. 2.5, which supports the existence of semantics in folksonomies. The next steps are a deeper understanding of the type of the relations hidden in folksonomies (cf. Sec. 4.3) and the development of methods to extract them (cf. Sec. 4.4).

4.2 Associations between Tags, Users, and Resources

As folksonomy systems grow larger, the users feel the need for more structure for better organizing their resources. For instance, approaches for tagging tags, or for bundling them, are discussed on the corresponding mailing lists e.g. the delicious-discuss list and are provided by some of the systems. A first step towards more structure within such systems is to discover knowledge that is already implicitly present by the way different users assign tags to resources. This knowledge may be used for recommending both a hierarchy of the already existing tags, and additional tags, ultimately leading towards emergent semantics by converging use of the same vocabulary.

[33] http://www.kde.cs.uni-kassel.de/ws/eswc2007/

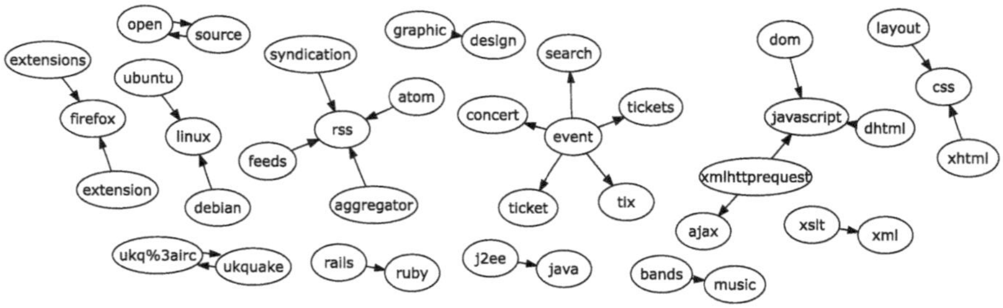

Fig. 3 Example of extracted association rules between tags

In [41], we focus on a certain KDD technique, namely association rules [1]. Since folksonomies provide a three-dimensional dataset (users, tags, and resources) instead of a usual two-dimensional one (items and transactions), we start in [41] with a systematic overview of projecting a folksonomy onto a two-dimensional structure. For one selected projection, we demonstrate here the outcome of association rule mining on a large-scale folksonomy dataset. The rules can be applied for different purposes, such as recommending tags, users, or resources, populating the supertag relation of the folksonomy, and community detection. Another example as well as details are described in [41].

To illustrate the outcome of the learning approach, an example from [41] is given in Figure 3. It shows all rules between tags from del.icio.us for a minimum support of 0.05% and a minimum confidence of 50%. In this example, rules of the form $A \to B$ can be read as "if a user has assigned the tag A to some resources, he often assigned tag B as well". If del.icio.us users are tagging some webpage with *debian*, they are likely to tag it with *linux* as well, and pages about *bands* are probably also tagged with *music*. As discussed in Section 4.1, we are looking for ways to discover subsumption relations which are needed to build ontologies, so that rule mining can be used to learn a taxonomic structure. As an example, consider the case where many resources tagged with *xslt* are also tagged with *xml*. This indicates that *xml* can be considered a supertopic of *xslt* if one wants to automatically populate the $\prec$ relation. Figure 3 also shows two pairs of tags which occur together very frequently without any distinct direction in the rule: *open source* occurs as a phrase most of the time, while the other pair consists of two tags (*ukquake* and *ukq:irc*), which seem to be added automatically to any resource that is mentioned in the chat channel ukq.

We can learn from these examples that it is possible to extract meaningful relations between tags from folksonomy data. To get a better understanding of what was extracted, we have to ground the extracted relations between the tags, users and resources by mapping them to an external knowledge source with a clear semantic meaning or better grounded relationships. We can try to do this for all three dimension, but we focus on tags, as they transport (most of) the semantic information. Such semantic information is captured by large lexical ontologies and thesauri, and we will use both to evaluate the meaning of different similarity measures between tags.

Therefore, in the next section, a fine grained analysis of various techniques we used here and in previous sections, namely, association rule mining and FolkRank ranking, is presented to further contribute to the understanding of the extracted relation by every method.

4.3 Understanding Tag Relatedness in Folksonomies

In this part we focus on the understanding of the specific relationship between tags in folksonomies. As we have seen in Sec. 2.5, the structure of folksonomies differs fundamentally from that of, e.g. natural text or web resources, and poses new challenges for the fields of knowledge discovery and ontology learning. Central to these tasks are the concepts of similarity and relatedness. In the previous section, among others, we introduced the computation of relations between tags by the association rule mining algorithm (based on co-occurrence) which can be easily turned into a tag relatedness measure. In [5], we focus on similarity and relatedness of tags, because they carry most of the semantic information within a folksonomy, and provide thus the link to ontologies and more formal semantics. Additionally, this focus allows for an evaluation with well-established measures of similarity in existing lexical databases.

Budanitsky and Hirst pointed out that similarity can be considered as a special case of relatedness [4]. As both similarity and relatedness are semantic notions, one way of defining them for a folksonomy is to map the tags to a thesaurus or lexicon like Roget's thesaurus[34] or WordNet [13], and to measure the relatedness there by means of well-known metrics. The other option is to define measures of relatedness directly on the network structure of the folksonomy. One important reason for using measures grounded in the folksonomy, instead of mapping tags to a thesaurus, is the observation that the vocabulary of folksonomies includes many community-specific terms which did not make it yet into any lexical resource. Measures of tag relatedness in a folksonomy can be defined in several ways. Most of these definitions use statistical information about different types of *co-occurrence* between tags, resources and users. Other approaches adopt the *distributional hypothesis* [14, 20], which states that words found in similar contexts tend to be semantically similar. This approach also retains the possibility to include "matured" folksonomy vocabulary back into the thesauri or lexicons, which addresses the inherent knowledge acquisition bottleneck problem of these systems. From a linguistic point of view, these two families of measures focus on orthogonal aspects of structural semiotics [10, 8].

---

[34] `http://www.gutenberg.org/etext/22`

The co-occurrence measures address the so-called syntagmatic relation, where words are considered related if they occur in the same part of text. The contextual measures address the paradigmatic relation (originally called associative relation by Saussure), where words are considered related if they can replace one another without affecting the structure of the sentence.

In most studies, the selected measures of relatedness seem to have been chosen in a rather ad-hoc fashion. We believe that a deeper insight into the semantic properties of relatedness measures is an important prerequisite for the design of ontology learning procedures that are capable of harvesting the emergent semantics of a folksonomy.

In [5], we analyse five measures of tag relatedness: the *co-occurrence count*, three *distributional measures* which use the cosine similarity [40] in the vector spaces spanned by users, tags, and resources, respectively, and *FolkRank* (cf. Sec. 3.2), our graph-based measure. Our analysis is based on data from a large-scale snapshot of the popular social bookmarking system del.icio.us.[35] To provide a semantic grounding of our folksonomy-based measures, we map the tags of del.icio.us to synsets of WordNet and use the semantic relations of WordNet to infer corresponding semantic relations in the folksonomy. In WordNet, we measure the similarity by using both the taxonomic path length and a similarity measure by Jiang and Conrath [29] that has been validated through user studies and applications [4]. The use of taxonomic path lengths, in particular, allows us to inspect the edge composition of paths leading from one tag to the corresponding related tags. This characterization proves to be especially insightful.

As a result, we show that distributional measures, which capture the context of a given tag in terms of resources, users, or other co-occurring tags, establish – in a statistical sense – *paradigmatic* relations between tags in a folksonomy. Strikingly, our analysis shows that the behavior of the most accurate measure of similarity (in terms of semantic distance of the indicated tags) can be matched by a computationally lighter measure (tag context similarity) which only uses co-occurrence with the popular tags of the folksonomy. In general, we show that a semantic characterization of similarity measures computed on a folksonomy is possible and insightful in terms of the type of relations that can be extracted. We show that despite a large degree of variability in the tags indicated by different similarity measures, it is possible to connotate *how* the indicated tags are related to the original one.

Another contribution of [5] addresses the question of emergent semantics: our results indicate clearly that, given an appropriate measure, globally meaningful tag relations can be harvested from an aggregated and uncontrolled folksonomy vocabulary. Specifically, we show that the measures based on tag and resource context are capable of identifying tags belonging to a common semantic concept. Admittedly, in their current status, none of the measures we studied can be seen as *the* way to instant ontology creation. However, we

[35] `http://del.icio.us/`

believe that further analysis of these and other measures, as well as research on how to combine them, will help to close the gap towards the Semantic Web.

Based on the results we have so far, the construction of tag hierarchies is the natural next step. We made a first attempt in [2], where we present results of a learned music style ontology. The data stems from last.fm[36], a music folksonomy system. A more advanced learning approach was applied on our del.icio.us dataset. The idea was to show that learning of ontologies from a large scale folksonomy is possible. In [45], an extended version of the algorithm from [2] is used for learning the ontology. The results are compared with Wordnet[37] and with the categorization scheme of Wikipedia.[38] Several drawbacks of the original algorithm could be solved and led to a better ontology. One central factor was the disambiguation of the word sense of polysemous tags and the calculation of synsets. Both approaches utilize the relatedness measures grounded before. While the synset detection algorithm reduces the number of tags by merging real synonyms as well as spelling variants, the word sense disambiguation component places tags more than once in the generated ontology. An example of a learned ontology is depicted in Fig. 4. As one can see, the tag *language* is placed under the tag *programming*, which hints the meaning of *language* in this case. We see the programming languages *lisp* as a sub-tag of *languages.* No names of natural languages like German are placed as sub-tags of *language* in this part of the graph. A more detailed description of the algorithm and the results can be found in [45].

4.4 Conceptual Structures in Folksonomies

Unlike ontologies, folksonomies do not suffer from the knowledge acquisition bottleneck, as the significant provision of content by many people shows. On the other hand, folksonomies – unlike ontologies [17] – do not explicitly state shared conceptualisations, nor do they force users to use tags consistently. However, the usage of tags of users with similar interests tends to converge to a shared vocabulary as explained in the previous section. Our intention is to discover these shared conceptualisations that are hidden in a folksonomy. To this end, we present in [30] an algorithm, TRIAS, for discovering subsets of folksonomy users who implicitly agree (on subsets of resources) on a common conceptualization.

Our algorithm returns a tri-ordered[39] set of triples, where each triple (A, B, C) consists of a set A of users, a set B of tags, and a set C of resources. These triples – called *tri-concepts* in the sequel – have the property that each user in A has tagged each resource in C with all tags from B, and

[36] `http://www.last.fm/`
[37] `http://wordnet.princeton.edu/`
[38] `http://www.wikipedia.org/`
[39] See [30] for details.

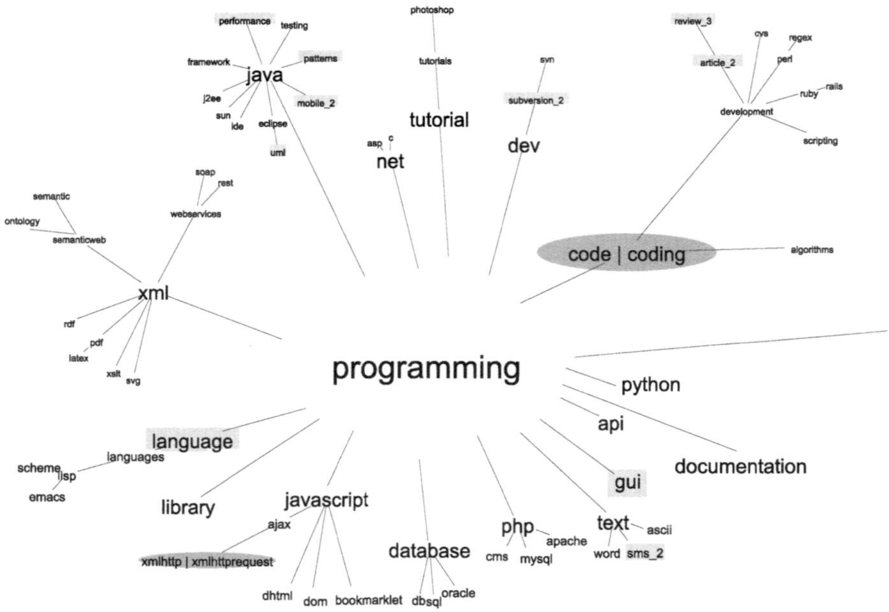

Fig. 4 Fragment of the learned ontology centered around programming. One meaning of language (programming languages) is depicted. (Figure is taken from [45])

that none of these sets can be extended without shrinking one of the other two dimensions. Each retrieved triple indicates thus a set A of users who (implicitly) share a conceptualisation, where the set B of tags is the intension of the concept, and the set C of resources is its extension. We can additionally impose minimum support constraints on each of the three dimensions "users", "tags", and "resources", to retrieve the most significant shared concepts only.

From a data mining perspective, the discovery of shared conceptualizations opens a new research field which may prove interesting also outside the folksonomy domain: "Closed itemset mining in triadic data", which is located on the confluence of the research areas of Association Rule Mining and Formal Concept Analysis.

In contrast to the already presented results of Sec. 4.3 and 4.2, TRIAS relates elements from different dimensions of the folksonomy. This allows for the simultaneous detection of hidden user groups and their interest expressed by the tags and the tagged resources. Another application could be the extraction of a concept hierarchy to learn ontologies as pointed out in [30].

The next step after discovering shared conceptualisations would be to formalize them in an ontology, and to combine and integrate this approach with the results of Sec. 4.3.

5 Conclusion and Future Work

Data Mining on folksonomies is a new research area attracting a lot of attention in the last years as new types of data with unknown and interesting properties appear. In this paper we presented the analysis of the properties of these new data, the application and adaption of known data mining approaches, and the usage of this data to extract semantic information. The three applications spam detection, ranking and recommendation were introduced and three approaches to extract the hidden semantic information from folksonomies were presented. Our own system BibSonomy was introduced as a platform where researchers manage their publication on a daily basis but also as a research environment to test new methods like ranking and recommendation which already found the way into the system.

In principle, the presented folksonomy mining approaches implement the ideas of Semantic Web Mining (cf. [47]). Therefore, they make our vision of utilizing mining to help to build the Semantic Web and to analyze it real. Hence, one long term goal is to use the weakly structured data of a folksonomy as data source for the Semantic Web. Further, convincing people to use a kind of "Semantic Bookmarking System" which is usable in the same easy way as the existing non-semantic versions is the vision and part of the future work. First steps in this direction with promising results were presented in this paper and we could show that it is possible to extract valuable information from folksonomies and to use data mining techniques to support user of social bookmarking systems.

A central phenomenon of the Web 2.0 is the contribution of many users distributed over the world but tied to a computer. The next step is to bring the web to mobile devices and to set up new services which do not only allow users to provide information but also to monitor their activities. This physical information will provide new kinds of data which allow for new services. A combination of the physical world with its small devices, sensors etc., the Web 2.0 look and feel, and the Semantic Web to connect everything will lead to the next generation of the Web.

Acknowledgements

This paper would not have been possible without the fruitful joint work with my colleagues from the KDE research group in Kassel, Germany, namely G. Stumme, C. Schmitz, R. Jäschke, D. Benz, and B. Krause. Thanks also to the collaborators A. Baldassarri, C. Cattuto, B. Ganter, V. Loreto, L. B. Marinho, F. Menczer, L. Schmidt-Thieme, and V. D. P. Servedio from outside the group. All papers summarized in this work are results of an inspiring collaboration with all of them.

Part of this research was funded by the European Union in the Nepomuk (FP6-027705) and Tagora (FET-IST-034721) projects.

References

1. Agrawal, R., Imielinski, T., Swami, A.: Mining association rules between sets of items in large databases. In: SIGMOD 1993: Proceedings of the 1993 ACM SIGMOD International Conference on Management of Data, pp. 207–216. ACM Press, New York (1993)
2. Benz, D., Hotho, A.: Position paper: Ontology learning from folksonomies. In: Hinneburg, A. (ed.) LWA 2007: Lernen - Wissen - Adaption, Halle, Workshop Proceedings (LWA), September 2007, pp. 109–112. Martin-Luther-University Halle-Wittenber (2007)
3. Brin, S., Page, L.: The Anatomy of a Large-Scale Hypertextual Web Search Engine. Computer Networks and ISDN Systems 30(1-7), 107–117 (1998)
4. Budanitsky, A., Hirst, G.: Evaluating WordNet-based measures of lexical semantic relatedness. Computational Linguistics 32(1), 13–47 (2006)
5. Cattuto, C., Benz, D., Hotho, A., Stumme, G.: Semantic grounding of tag relatedness in social bookmarking systems. In: Sheth, A.P., Staab, S., Dean, M., Paolucci, M., Maynard, D., Finin, T., Thirunarayan, K. (eds.) ISWC 2008. LNCS, vol. 5318, pp. 615–631. Springer, Heidelberg (2008)
6. Cattuto, C., Loreto, V., Pietronero, L.: Collaborative tagging and semiotic dynamics, arXiv:cs.CY/0605015 (May 2006)
7. Cattuto, C., Schmitz, C., Baldassarri, A., Servedio, V.D.P., Loreto, V., Hotho, A., Grahl, M., Stumme, G.: Network properties of folksonomies. AI Communications 20(4), 245–262 (2007)
8. Chandler, D.: Semiotics: The Basics, 2nd edn. Taylor & Francis, Abington (2007)
9. Cimiano, P., Hotho, A., Staab, S.: Learning Concept Hierarchies from Text Corpora using Formal Concept Analysis. Journal of Artificial Intelligence Research (JAIR) 24, 305–339 (2005)
10. de Saussure, F.: Course in General Linguistics. Duckworth, London [1916] (1983) (trans. Roy Harris)
11. Dubinko, M., Kumar, R., Magnani, J., Novak, J., Raghavan, P., Tomkins, A.: Visualizing tags over time. In: Proceedings of the 15th International WWW Conference (May 2006)
12. Fayyad, U.M., Piatetsky-Shapiro, G., Smyth, P.: From data mining to knowledge discovery: An overview. In: Advances in Knowledge Discovery and Data Mining, pp. 1–34. MIT Press, Cambridge (1996)
13. Fellbaum, C. (ed.): WordNet: an electronic lexical database. MIT Press, Cambridge (1998)
14. Firth, J.R.: A synopsis of linguistic theory 1930-55. Studies in Linguistic Analysis (special volume of the Philological Society) 1952-59, 1–32 (1957)
15. Ganter, B., Wille, R.: Formal Concept Analysis: Mathematical Foundations. Springer, Heidelberg (1999)
16. Golder, S., Huberman, B.A.: The structure of collaborative tagging systems. Journal of Information Science 32(2), 198–208 (2006)
17. Gruber, T.R.: Towards Principles for the Design of Ontologies Used for Knowledge Sharing. In: Guarino, N., Poli, R. (eds.) Formal Ontology in Conceptual Analysis and Knowledge Representation, Deventer, Netherlands. Kluwer, Dordrecht (1993)
18. Halpin, H., Robu, V., Shepard, H.: The dynamics and semantics of collaborative tagging. In: Proceedings of the 1st Semantic Authoring and Annotation Workshop (SAAW 2006), vol. 209. CEUR-WS (2006)

19. Hammond, T., Hannay, T., Lund, B., Scott, J.: Social Bookmarking Tools (I): A General Review. D-Lib Magazine 11(4) (April 2005)
20. Harris, Z.S.: Mathematical Structures of Language. Wiley, New York (1968)
21. Heymann, P., Koutrika, G., Garcia-Molina, H.: Fighting spam on social web sites: A survey of approaches and future challenges. IEEE Internet Computing 11(6), 36–45 (2007)
22. Hotho, A.: Social bookmarking. In: Back, A., Gronau, N., Tochtermann, K. (eds.) Web 2.0 in der Unternehmenspraxis: Grundlagen, Fallstudien und Trends zum Einsatz von Social Software, pp. 26–38. Oldenbourg Verlag, München (2008)
23. Hotho, A., Benz, D., Jäschke, R., Krause, B., (eds.): ECML PKDD Discovery Challenge 2008 (RSDC 2008). Workshop at 18th Europ. Conf. on Machine Learning (ECML 2008) / 11th Europ. Conf. on Principles and Practice of Knowledge Discovery in Databases, PKDD 2008 (2008)
24. Hotho, A., Jäschke, R., Schmitz, C., Stumme, G.: BibSonomy: A social bookmark and publication sharing system. In: Proceedings of the Conceptual Structures Tool Interoperability Workshop at the 14th International Conference on Conceptual Structures, pp. 87–102. Aalborg University Press, Aalborg (2006)
25. Hotho, A., Jäschke, R., Schmitz, C., Stumme, G.: Information retrieval in folksonomies: Search and ranking. In: Sure, Y., Domingue, J. (eds.) ESWC 2006. LNCS, vol. 4011, pp. 411–426. Springer, Heidelberg (2006)
26. Hotho, A., Jäschke, R., Schmitz, C., Stumme, G.: Trend detection in folksonomies. In: Avrithis, Y., Kompatsiaris, Y., Staab, S., O'Connor, N.E. (eds.) SAMT 2006. LNCS, vol. 4306, pp. 56–70. Springer, Heidelberg (2006)
27. Illig, J.: Machine learnability analysis of textclassifications in a social bookmarking folksonomy. Bachelor thesis, University of Kassel, Supervisor: Andreas Hotho, Kassel (2008)
28. Illig, J., Hotho, A., Jäschke, R., Stumme, G.: A comparison of content-based tag recommendations in folksonomy systems. In: Postproceedings of the International Conference on Knowledge Processing in Practice (KPP 2007). Springer, Heidelberg (2009) (to appear)
29. Jiang, J.J., Conrath, D.W.: Semantic similarity based on corpus statistics and lexical taxonomy. CoRR, cmp-lg/9709008 (1997)
30. Jäschke, R., Hotho, A., Schmitz, C., Ganter, B., Stumme, G.: Discovering shared conceptualizations in folksonomies. Web Semantics: Science, Services and Agents on the World Wide Web 6(1), 38–53 (2008)
31. Jäschke, R., Marinho, L., Hotho, A., Schmidt-Thieme, L., Stumme, G.: Tag recommendations in social bookmarking systems. AI Communications 21(4), 231–247 (2008)
32. Jäschke, R., Marinho, L.B., Hotho, A., Schmidt-Thieme, L., Stumme, G.: Tag recommendations in folksonomies. In: Kok, J.N., Koronacki, J., Lopez de Mantaras, R., Matwin, S., Mladenič, D., Skowron, A. (eds.) PKDD 2007. LNCS (LNAI), vol. 4702, pp. 506–514. Springer, Heidelberg (2007)
33. Kosala, R., Blockeel, H.: Web mining research: A survey. SIGKDD Explorations 2(1), 1–15 (2000)
34. Krause, B., Schmitz, C., Hotho, A., Stumme, G.: The anti-social tagger - detecting spam in social bookmarking systems. In: Proc. of the Fourth International Workshop on Adversarial Information Retrieval on the Web, pp. 61–68. ACM, New York (2008)

35. Lehmann, F., Wille, R.: A triadic approach to formal concept analysis. In: Ellis, G., Rich, W., Levinson, R., Sowa, J.F. (eds.) ICCS 1995. LNCS, vol. 954, pp. 32–43. Springer, Heidelberg (1995)
36. Lund, B., Hammond, T., Flack, M., Hannay, T.: Social Bookmarking Tools (II): A Case Study - Connotea. D-Lib Magazine 11(4) (April 2005)
37. Mathes, A.: Folksonomies – Cooperative Classification and Communication Through Shared Metadata (December 2004), `http://www.adammathes.com/academic/computer-mediated-communication/folksonomies.html`
38. Mika, P.: Ontologies Are Us: A Unified Model of Social Networks and Semantics. In: Gil, Y., Motta, E., Benjamins, V.R., Musen, M.A. (eds.) ISWC 2005. LNCS, vol. 3729, pp. 522–536. Springer, Heidelberg (2005)
39. Patashnik, O.: BibTeXing (Included in the BIBTEX distribution) (1988)
40. Salton, G.: Automatic text processing: the transformation, analysis, and retrieval of information by computer. Addison-Wesley Longman Publishing Co. Inc., Boston (1989)
41. Schmitz, C., Hotho, A., Jäschke, R., Stumme, G.: Mining association rules in folksonomies. In: Batagelj, V., Bock, H.-H., Ferligoj, A., Ziberna, A. (eds.) Data Science and Classification (Proc. IFCS 2006 Conference) Studies in Classification, Data Analysis, and Knowledge Organization, pp. 261–270. Springer, Heidelberg (2006)
42. Staab, S., Santini, S., Nack, F., Steels, L., Maedche, A.: Emergent semantics. Intelligent Systems, IEEE [see also IEEE Expert] 17(1), 78–86 (2002)
43. Staab, S., Studer, R. (eds.): Handbook on Ontologies. International Handbooks on Information Systems. Springer, Heidelberg (2004)
44. Steels, L.: The origins of ontologies and communication conventions in multi-agent systems. Autonomous Agents and Multi-Agent Systems 1(2), 169–194 (1998)
45. Stützer, S.: Lernen von Ontologien aus kollaborativen Tagging-Systemen. Master thesis, University of Kassel, Supervisor: Andreas Hotho, Kassel (2009)
46. Stumme, G.: A finite state model for on-line analytical processing in triadic contexts. In: Ganter, B., Godin, R. (eds.) ICFCA 2005. LNCS (LNAI), vol. 3403, pp. 315–328. Springer, Heidelberg (2005)
47. Stumme, G., Hotho, A., Berendt, B.: Semantic web mining - state of the art and future directions. Journal of Web Semantics 4(2), 124–143 (2006)
48. Tonkin, E., Guy, M.: Folksonomies: Tidying up tags? D-Lib 12(1) (2006)
49. Wetzker, R., Umbrath, W., Said, A.: A hybrid approach to item recommendation in folksonomies. In: ESAIR 2009: Proceedings of the WSDM 2009 Workshop on Exploiting Semantic Annotations in Information Retrieval, pp. 25–29. ACM, New York (2009)
50. Wille, R.: Restructuring lattice theory: an approach based on hierarchies of concepts. In: Rival, I. (ed.) Ordered sets, pp. 445–470, Reidel (1982)

A Uniform Theoretic Approach to Opinion and Information Retrieval

G. Amati, G. Amodeo, M. Bianchi, C. Gaibisso, and G. Gambosi

Abstract. In this paper, we introduce a supervised method for the generation of a dictionary of weighted opinion bearing terms from a collection of opinionated documents. We also describe how such a dictionary is used in the framework of an algorithm for opinion retrieval, that is for the problem of identifying the documents in a collection where some opinion is expressed with respect to a given query topic. Several experiments, performed on the TREC Blog collection, are reported together with their results; in these experiments, the use of different combinations of DFR (Divergence from Randomness) probabilistic models to assign weights to terms in the dictionary and to documents is studied and evaluated. The results show the stability of the method and its practical utility. Moreover, we investigate the composition of the generated lexicons, mainly focusing on the presence of stop-words. Quite surprisingly, the best performing dictionaries show a predominant presence of stop-words. Finally, we study the effectiveness of the same approach to generate dictionaries of polarity-bearing terms: preliminary results are provided.

1 Introduction and Related Works

Sentiment analysis is a type of text classification, where text is classified by types of opinions, sentiments, or, more generally, by the subjectivity contained in the

Giambattista Amati
Fondazione Ugo Bordoni, Rome, Italy
e-mail: gba@fub.it

Giuseppe Amodeo
Dept. of Computer Science, University of L'Aquila, Italy
e-mail: gamodeo@fub.it

Marco Bianchi · Carlo Gaibisso
Istituto di Analisi dei Sistemi ed Informatica "Antonio Ruberti" - CNR, Rome, Italy
e-mail: surname@iasi.cnr.it

Giorgio Gambosi
Dept. of Mathematics, University of Rome "Tor Vergata", Italy
e-mail: gambosi@mat.uniroma2.it

G. Armano et al. (Eds.): Intelligent Information Access, SCI 301, pp. 83–108.
springerlink.com

text. The study and evaluation of efficient solutions to detect sentiments in text is a popular research area, and techniques are applied coming from natural language processing, computational linguistics, machine learning, information retrieval and text mining.

In the simplest case, text can be classified just by considering the presence of subjective or opinionated sentences, irrespective of their *polarity*, that is not taking into account whether documents contain positive or negative opinions. Additionally, analysis can go further and a polarity degree of opinions can be associated to text.

As a classification problem sentiment analysis can be handled by either a supervised or unsupervised method. A general introduction to classification techniques for sentiment analysis can be found in [20] and [32]. Supervised techniques attempt to train a sentiment classifier on the basis of the frequencies of words in the documents. Several papers, such as [33] and [40], indicate that standard machine learning methods perform very well.

Unsupervised techniques (also referred as lexicon-based methods) derive lexicons of "positive" and "negative" terms, and then compute an overall attitude score for a document on the basis of the occurrence of such terms. For example, if a document contains more positive than negative terms it is deemed as positive, else it is assigned as negative.

Since unsupervised techniques do not require any training phase, they are best suited to the case when no training data are available. Relevant case studies are presented in [11] and [41]. However, prediction accuracy my be affected by the quality of the underlying linguistic resources or by a correlation between specific topics and the used external resources.

The fusion of search by content and classification by sentiment, known as *Opinion Retrieval* or *Opinion Finding*, is now a hot and prolific research area in Information Retrieval.

In addition to the usual sentiment classification problem, opinion retrieval also requires that documents need to be predicted as relevant with respect to a given topic.

Since 2006, the Text REtrieval Conference (TREC) has an evaluation track on Blogs where the main task is Opinion Retrieval, that is the task of selecting the opinionated blog posts relevant to a given topic from a collection made of more than 3.2 million blog Web pages [24, 29, 31]. The opinion finding task is thus a specification of the search task, when the user is trying to discover on the blogosphere (or, in general, in a document corpus) the drift in public opinion about a given named-entity target [30], such as products or people.

The Blog collection [23] is the largest and the only, or at least the most reliable, publicly available data set for the evaluation of opinion retrieval techniques. In 2008, the topic data set consists of 150 information topics together with the list of relevant and opinionated assessed documents.

As reported in [30], the opinion finding is normally approached by as a two-stage process. In the first stage, topic-relevant documents are ranked using a retrieval model. In the second stage, the retrieved documents are re-ranked taking into account opinion finding features, often through a combination of the first stage retrieval score with a computed score denoting the degree of opinionated content in the document (i.e. opinion score). One of the most effective approaches to evaluate the degree of opinionated content is to submit the dictionary of opinion bearing terms as a standard query, and thus to compute an "opinion relevance" score of the documents. Many variants of this approach have been proposed: the dictionary can be automatically or manually built, the terms can be automatically or manually weighted, distances between terms of the topic and opinion bearing terms can be considered or not. An overview of the latest approaches to opinion finding can be found in [31].

In this paper, we report the experimentation and the evaluation of different variants of a supervised method for the automatic generation of a dictionary of weighted opinion bearing terms from a collection of opinionated documents. Different combinations of DFR (Divergence from Randomness) probabilistic models are applied to assign weights to terms in the lexicons and to documents and their effectiveness evaluated. We also investigate the composition of the generated lexicons, mainly for what regards the presence of stop-words. Quite surprisingly, the best performing lexicons are the ones showing a predominant presence of stop-words. Finally, we checked whether effective lexicons of polarity-bearing terms can be generated by the same approach: preliminary results are shown.

The paper is organized as follows: sections 2 and 3 introduce and motivate the statistical foundations of our approach to the automatic construction of lexicons and their adoption in opinion retrieval, respectively. In section 4 our main experimentation settings are listed and motivated. In section 5 the results of the experiments are presented and discussed. In section 6 the applicability to polarity of the opinion retrieval approach is discussed. In Section 7 we present the conclusions.

2 Automatic Construction of a Sentimental Lexicon

In this section we introduce and motivate the statistical foundations of the automatic construction of a lexicon of weighted opinion-bearing terms to be used in opinion retrieval, first introduced in [4].

The statistical approach to Information Retrieval of Divergence From Randomness (DFR) probabilistic models [6, 1] can be used for Opinion Retrieval. The DFR models are based on a simple idea, that is that the higher the divergence of within-document term-frequency is from the within-the-collection frequency, the higher the information of the term in the document is.

Before introducing a DFR model, we provide some notations. Let:

- C be the collection;
- $d \in C$ be a generic document, considered as a multiset of terms (bag of words);
- $l(d)$ be the number of tokens of the terms in d (the *document length*);
- $R \subseteq C$ be the set of relevant documents of C, with respect to a set T of topics;
- $O \subseteq R$ be the set of relevant and opinionated documents of C, with respect to T;
- V be the set of terms occurring in C, the *term lexicon*;
- $t \in V$ be a generic term;

- $tf_{t,d}$ be the frequency of t in d;
- $TF_{t,D} = \sum_{d \in D} tf_{t,d}$, be the frequency of t in a subset D of C;
- $DF_{t,D}$ be the number of documents of D in which t occurs (the *document frequency*);
- $Pr_{t,d} = \frac{tf_{t,d}}{l(d)}$, be the relative frequency of t in d;
- $Pr_{t,D} = \frac{TF_{t,D}}{\sum_{d \in D} l(d)}$, be the relative frequency of t in the set of documents in D.

First, we define the *information content* of a term and then an DFR model. The information content of a term t on a document d is given by:

$$\text{Inf}(t) = -\log_2 \Pr(Pr_{t,d} | Pr_{t,C}, l(d)) \tag{1}$$

A DFR model then specifies:

- how to compute in Equation 1 the probability Pr of observing the frequency $Pr_{t,d}$ in a document of length $l(d)$ given a frequency $Pr_{t,C}$,
- how to normalize the information content by taking into account the size $l(d)$ of the considered sample d.

A fundamental property of a DFR model is that, if the posterior probability $Pr_{t,d}$ of a term, that is the frequency observed in a subset of documents or in a single document d of length $l(d)$, equals the prior, that is the frequency $Pr_{t,C}$ with respect to the collection, then the information content is minimal. As opposite, when there is a high deviation of $Pr_{t,d}$ from the prior $Pr_{t,C}$, then the information content is equally high.

The probability deviation property explains why the name of divergence from randomness models: terms that occur randomly in text have a low divergence of the two probabilities $Pr_{t,d}$ and $Pr_{t,C}$. For example, stop-words are terms for which the observed and expected frequencies coincide more than for information bearing words. When $Pr_{t,d} \sim Pr_{t,C}$, the term t does not bring relevant information to the document d. On the other hand, the more the deviation of $Pr_{t,d}$ is from $Pr_{t,C}$, the more the information is carried by t in d.

Unfortunately, from a information theoretic point of view, *opinion terms* should not bring information being terms that appear randomly and independently from the content and, consequently, $\text{Inf}(t)$ should be always low. Fortunately, we can learn from Blog data sets and make statistical analysis on the divergence of the distributions of the terms in the set of opinionated and relevant documents O from the superset of relevant (with or without opinions) documents R. We then may use a DFR model by learning on the divergence of the term-frequency in the set of opinionated and relevant documents from the term-frequency in the set of relevant documents only.

We now introduce two information theoretic functions, on which one of the DFR models is based, and also provide approximations to such functions, which will be used to efficiently build the dictionaries.

To identify terms that are opinion-bearing, let us first of all define a measure $OE(\cdot)$, the *opinion entropy function*, for the opinion content of a term. In doing this, we argue that opinion terms tend to appear more frequently in O than in R.

$OE(\cdot)$ assigns a weight to t according to the strength of the opinion it bears:

$$OE(t) = -\log_2 \Pr(Pr_{t,O}|Pr_{t,R}). \tag{2}$$

More precisely, $OE(\cdot)$ measures the average information t brings on O. In other words, it quantifies how much a term is representative of O as a function of the relative frequency in O and R. Hence, terms with an high $OE(\cdot)$ are candidate members of our lexicon.

Note that, according to the definition of $OE(\cdot)$, content-bearing terms tend to have a similar relative frequencies in both R and O, and consequently, they have $OE(\cdot) \sim 0$.

To compute the opinion entropy values for terms in the collection, we use an approximation of equation 2: the Kullback-Leibler divergence [17, 18]. In probability and information theory, the Kullback-Leibler (KL) divergence is a non-commutative measure[1] of the difference between the probability distributions of two discrete random variables P and Q, defined on the same set of events Ω. The KL divergence of Q from P is defined as:

$$KL(P||Q) = \sum_{\omega \in \Omega} P(\omega) \cdot \log_2 \frac{P(\omega)}{Q(\omega)} \tag{3}$$

Here, we have a simple probabilistic space containing two events: the opinionated and relevant set and the relevant set. In this case, the opinion entropy function of Equation 5 can be approximated because:

$$KL(Pr_{t,O}||Pr_{t,R}) \sim \frac{-\log_2 \Pr(Pr_{t,O}|Pr_{t,R})}{\sum_{t \in V} TF_{t,O}} \tag{4}$$

where $\sum_{t \in V} TF_{t,O}$ is the total number of terms in O. Being this sum common to all terms, it can be omitted in the opinion entropy estimation. Being $Pr_{t,O} > Pr_{t,R}$, resulting that

$$OE(t) = -\log_2 \Pr(Pr_{t,O}|Pr_{t,R}) \propto KL(Pr_{t,O}||Pr_{t,R}) \tag{5}$$

up to a proportional factor and a small error. In short, to identify and to weight terms of our lexicons, we use KL divergence as approximation of $OE(\cdot)$.

Note that the $OE(\cdot)$ function has been empirically shown to work properly, but with some relevant exceptions:

- rare content-bearing terms occurring in both R and O. For example, if one of these terms occurs 3 times in O and 4 times in R, then $OE(\cdot)$ is improperly high;

[1] Although KL divergence is often seen as a distance metric, it is not symmetric and does not satisfy the triangle inequality (hence 'divergence' rather than 'distance').

- a topic has a large number of relevant documents that are also opinionated. In such a case, it may happen that a term related to a topic has a moderate frequency in O because it has a high frequency in the unbalanced set of opinionated and relevant documents for that topic. But since $OE(\cdot)$ is computed with respect to the whole set of topics of interest, these content-bearing terms could be identified as opinionated.

Since KL does not consider if the term frequency is uniformly distributed across all opinionated documents or the term appears more densely in few documents, approximating $OE(\cdot)$ by KL does not avoid noisy terms to appear in the lexicons.

To deal with this improper behavior, let us introduce the *average opinion entropy*, which is used to filter out noisy terms:

$$AOE(t) = -\frac{1}{|O|}\sum_{d \in O} \log_2 \Pr(Pr_{t,d}|Pr_{t,O}). \tag{6}$$

$AOE(\cdot)$ measures the average divergence of term t document frequency from the expected term frequency of t in O. As stated in [4], noisy terms have low values of $AOE(\cdot)$, as $Pr_{t,d} \sim Pr_{t,O}$ for most opinionated documents. To filter these noisy terms it is thus enough to filter terms with low values of $AOE(\cdot)$.

Also for the $AOE(\cdot)$ function we introduce an approximation. Let us define the set

$$Lex_k = \{t \mid DF_{t,O} \geq k\} \tag{7}$$

of all the terms that appear in at least k documents of O. As shown in [4], in order to minimize $AOE(\cdot)$ for the set of candidates terms, i.e. the terms with the highest value of $OE(\cdot)$, we have to maximize $DF_{t,O}$.

Intuitively, the higher is the number of documents containing a term t, the higher is the probability that t is an opinion-bearing term. At the same time, the larger is k, the smaller is Lex_k: if k is too high, opinion bearing-terms could be filtered out. In practice, the optimal value of k is determined with the aim of obtaining a low enough lexicon size, without reducing the effectiveness of the retrieval.

Formally, we can approximate also $AOE(t)$ using $KL(Pr_{t,d}||Pr_{t,O})$, that is:

$$AOE(t) \sim \frac{1}{|O|}\sum_{d \in O} KL(Pr_{t,d}||Pr_{t,O}) = \frac{1}{|O|}\sum_{d \in O} Pr_{t,d} \cdot \log_2 \frac{Pr_{t,d}}{Pr_{t,O}} \tag{8}$$

Since O is a large sample of the collection C, we can approximate also the prior probability $Pr_{t,O}$ as follows:

$$Pr_{t,O} = \frac{TF_{t,O}}{\sum_{t \in V} TF_{t,O}} \sim \frac{TF_{t,O}}{|O| \cdot \overline{l(d)}} \tag{9}$$

where $\overline{l(d)}$ is the average documents length in O.

Remind that opinionated terms do not carry information content, in our assumption: opinion-bearing terms distribute more uniformly in the set of opinionated documents. This means that $Pr_{t,d} \sim Pr_{t,O}$ or, more generally, KL divergence is minimized.

So under the hypothesis that t distributes uniformly, we can define the posterior probability as:

$$Pr_{t,d} = \frac{tf_{t,d}}{l(d)} \sim \frac{TF_{t,O}}{DF_{t,O} \cdot \overline{l(d)}} \tag{10}$$

where $DF_{t,O}$ is the number of opinionated documents containing the t.

As a consequence, to find opinionated terms means to find those terms that *minimize*:

$$AOE(t) \propto - \sum_{d \in O} \frac{TF_{t,O}}{DF_{t,O}} \log_2 DF_{t,O} = -DF_{t,O} \cdot \frac{TF_{t,O}}{DF_{t,O}} \log_2 DF_{t,O} = -TF_{t,O} \cdot \log_2 DF_{t,O} \tag{11}$$

Since the approximating expression is negative, and since we may suppose that all terms have a frequency $TF_{t,O}$ of a similar order of magnitude in the set of opinionated documents, we may instead *maximize* the function

$$\log_2 DF_{t,O} \propto DF_{t,O} \tag{12}$$

Therefore the higher is the number of documents containing a term, the higher is the probability that the term is opinionated. Since we use a fast and effective implementation of the AOE function, that is the minimal number k of opinionated and relevant documents containing candidate terms, a sequence of weighted dictionaries Lex_k are built at different level of k.

Thus the $OE(\cdot)$ and $AOE(\cdot)$ functions can be adopted in order to construct an opinion lexicon as follows:

1. terms appearing in all the documents of O are weighted by $OE(\cdot)$, determining the candidates to appear in the lexicon as terms that achieved the highest $OE(\cdot)$ values;
2. only candidates with the lowest value of $AOE(\cdot)$ are selected for the lexicon.

For the sake of efficiency, we do not compute the exact values of $OE(\cdot)$ and $AOE(\cdot)$, but instead we approximate them as shown.

3 A Lightweight Opinion Retrieval Algorithm

Opinion retrieval for a given topic is typically accomplished in two steps [29, 24, 31]:

1. all the documents relevant with respect to the topic are retrieved, weighted, and ranked;
2. documents in the resulting ranking are re-ranked on the basis of the opinion strength they express.

More formally, $\Pr(q|d)$ measures the relevance of d to q. In the statistical approach to IR, this probability is proportional to the score of relevance assigned to the document with respect to the query, i.e.

$$\Pr(q|d) \propto Score_t(d,q) \tag{13}$$

Analogously, a second probability distribution is associated to the process of determining the opinion content of a document. If V is the sentimental dictionary

$$\Pr(V|d) \propto Score_o(d) \tag{14}$$

which is proportional to the score of opinion assigned to the document.

Moreover, following the usual assumption that document relevance is distributed according to Zipf law [43], we assert that:

$$\Pr(Z|d,\alpha) \propto \frac{k_Z}{r_Z(d)^\alpha} \tag{15}$$

where Z is a random variable modeling some kind of relevance (either with respect to a topic q or to express an opinion from V), k_Z is a constant and $r_Z(d)$ is the rank of d induced by the scoring function associated to Z, with parameter $\alpha \geq 1$.

Notice that in opinion retrieval we are interested in the joint probability $\Pr(q,V \mid d)$ of q and V.

Finally, we assume that the presence of opinion expressions in a document is not related to its topic relevance, i.e. if a document is relevant for a topic, the probability that it is opinionated is not affected by the topic relevance, and vice versa. That is, we assume that q and V are independent random variables.

Hence we have:

$$\Pr(q,V \mid d) = \Pr(q \mid d) \cdot \Pr(V \mid d) \propto \frac{Score_o(d)}{r_X(d,q)} \tag{16}$$

or equally

$$\Pr(q,V \mid d) = \Pr(q \mid d) \cdot \Pr(V \mid d) \propto \frac{Score_t(d,q)}{r_Y(d)} \tag{17}$$

where $r_X(d,q)$ is the *content rank* for all documents according to $Score_t(d,q)$ and $r_Y(d)$ is the *opinion rank* for all documents according to $Score_o(d)$.

These approximations are the basis of our three steps re-ranking algorithm:

1. given a query q related to a topic, the *content score* of the documents is assigned by means of a term-document matching function:

$$Score_t(d,q) \tag{18}$$

   The content rank $r_X(d,q)$ is then derived from $Score_t(d,q)$.
2. an *opinion score* is assigned to each retrieved document. This score is computed submitting the entire dictionary Lex_k as a query:

$$Score_o(d,Lex_k) \tag{19}$$

The *opinion score* with respect to q is defined as follow[2]:

$$Score_o(d,q) = \frac{Score_o(d,Lex_k)}{r_X(d)} \tag{20}$$

A new opinion rank $r^*_Y(d)$ is then derived from $Score_o(d,q)$.

3. Document ranking is boosted applying the dual function of $Score_o(d,q)$:

$$Score^+(d,q) = \frac{Score_t(d,q)}{r^*_Y(d)} \tag{21}$$

The final opinion ranking is obtained re-ranking the documents by $Score^+(d,q)$.

In our experimentation we initially tested a single re-ranking process using separately the approximations proposed by equations 16 and 17 to compute an opinion score. The performances of the two resulting rankings are very close. With found instead a meaningful increment using both of these, so defining our algorithm with a double re-ranking according to the equations 20 and 21.

4 Experimentation Goals and Motivations

Our main experimentation goal is to test several different aspects of the method described above to automatically generate lexicons of opinion-bearing terms. More in details, we are interested in verifying:

- the stability of the method, i.e. the independence of its effectiveness from the choice of the training set;
- the effectiveness of the re-ranking process, measured as the mean *MAP* variation introduced with respect to our baselines;
- the practical utility of the generated lexicons: too big lexicons could in fact make the process of document weighting unacceptably time expensive;
- the effectiveness of *KL* in expanding lexicons and weighting their terms. In particular, we will compare *KL* with *Bo1*, using the Bose-Einstein statistics, which is considered one of the most effective DFR term weighting models [34, 22]

We also aim to select and evaluate an effective DFR model, to perform a better weighting of the strength of the opinion expressed by documents. In order to do that, we compared the *DFree1* and the *DPH* models [4]. We focused ourselves on

[2] Note that a relevance ranking in general is a mixture of a normal distribution for relevant documents and an exponential distribution for non-relevant documents [25]. Since for any query the non relevant documents are the large majority of the documents of the collection, ranking roughly follows the power law, that is the probability of relevance of a document is inversely proportional to its document rank. Therefore:

$$Score_o(d,q) \propto Score_o(d,Lex_k) \cdot \Pr(d,q)$$

these alternatives since: *Dfree1* [2] has been empirically proven to be less sensitive to the specific characteristics of the data set; the results collected in our participation to the TREC 2007 blog track [3] showed that, among the tested models, *DPH* best performed with a collection of blog posts.

Finally, we are also interested in investigating the contents of the generated lexicons, mainly focusing on the presence of stop-words, those usually discarded from documents to effectively implement a topic retrieval.

This experimentation extends in a considerable way the one reported in [4].

In this section the experimentation environment, by which our results have been collected, is described in details. The environment consists of a standard test collection and an open source information retrieval framework, which are described in the following subsections.

4.1 The Opinion Retrieval Test Collection

According to [26], to measure the effectiveness of our proposals in a standard way we need a test collection made by:

- a blog posts collection;
- a test suite of information needs (topics);
- a relevance/opinion judgment for the blog posts in the collection.

4.1.1 The Blog Posts Collection

The collection we used is the TREC Blog Collection (*BLOGS06*) [29], a TREC test collection created and distributed by the University of Glasgow. *BLOGS06* contains blogs pages and was crawled over a period of 11 weeks, from December 2005 to February 2006. The total size of the collection amounts to 148 GB with three main different components: feeds (38.6 GB), permalinks (88.8GB), and homepages (20.8 GB). The collection contains spam as well as possibly non-blogs and non-English pages. In what follows the main characteristics of the collection are listed in a systematic way:

- Feeds:
  - Total Number of Feeds: 100,649
  - Total Number of Feeds collected: 753,681
  - Average feeds collected every day: 10,615
  - Uncompressed Size: 38.6GB
  - Compressed Size: 8.0GB
- Permalink Documents:
  - Total number of permalink documents: 3,215,171
  - Average documents every day: 45,284
  - Uncompressed Size: 88.8GB
  - Compressed Size: 12.6GB

- Homepage Documents:
  - Total number of homepage documents: 324,880
  - Average homepage documents collected every day: 4,576
  - Uncompressed Size: 20.8GB
  - Compressed Size: 4.0GB

Only the permalink component of the collection has been considered in the experimentation, denoted as *Blogs* in what follows. *Blogs* consists of 3.2 millions of Web pages, each one containing a post and the related comments. For the sake of simplicity, *blog post*, *permalink* and *document* will be interchangeably used in following sections.

4.1.2 Topics and Queries

There are many different types of blogs, some concerning a specific topic, some covering several ones, and others talking about personal daily life. In the latest three years, NIST identified 150 topics [29, 24, 31] of interest for blog retrieval. These topics were selected from a donated collection of queries sent to commercial blog search engines over the time period that the *BLOGS06* was collected. NIST assessors mainly created the topics by selecting queries, and building topics around those queries. Each topic follows the number, title, description, and narrative structure, shown by the following example:

```
<num> Number: 851
<title> "March of the Penguins"

<desc> Description:

Provide opinion of the film documentary "March of
the Penguins".

<narr> Narrative:

Relevant documents should include opinions concerning the
film documentary "March of the Penguins".  Articles or
comments about penguins outside the context of this film
documentary are not relevant.
```

We extracted a query from each topic, consisting in the content of the *Title* field. For the sake of simplicity, in what follows *query* and *topic* will be interchangeably used. Let *Topics* denote the set of provided topics.

4.1.3 The Set of Relevance/Opinion Judgments

NIST also provided a relevance/sentiment judgment file, named *Qrels*. Each entry in the file identifies a topic t, a document d and a judgment j as shown by the following example.

```
  t                    d                    j

901 ... BLOG06-20051206-025-0029297277 1
901 ... BLOG06-20051206-051-0007821418 0
901 ... BLOG06-20051208-027-0000033998 3
                    ...
901 ... BLOG06-20051213-023-0007797892 4
920 ... BLOG06-20060109-006-0009964253 2
```

The judgment values of relevance/sentiment are assigned as follows:

- 0: *d* is not relevant with respect to *t*;
- 1: *d* is relevant with respect to *t*, but does not comment upon it;
- 2: *d* is relevant with respect to *t* and positively comments upon it;
- 3: *d* is relevant with respect to *t* and neutrally comments upon it;
- 4: *d* is relevant with respect to *t* and negatively comments upon it;

It is worth to notice that the order in which the documents appear in the file is not indicative of the degree of relevance/sentiment. Finally, documents assumed to be irrelevant are not listed in the *Qrels*.

4.1.4 The Baselines

To verify the effectiveness of our lexicon based approach it will be compared to a standard IR system not supporting any opinion-finding features. To this end, five additional runs, named *baselines*, produced using standard off-the-shelf IR systems, denoted by $BL_1, BL_2, \ldots, BL_5$ and provided by NIST have been considered for the comparison. Each entry in a baseline, at least for what regards our interest, identifies a topic *t*, a document *d* related to *t*, the rank *r* of *d*, and the score *s* assigned to *d* according to which the ranking has been established, as shown in the following example:

```
 t                    d                    r         s

851 ... BLOG06-20051206-025-0019312838 0   7.923441282439574 ...
851 ... BLOG06-20051206-050-0020317610 1   6.825848636554143 ...
851 ... BLOG06-20051206-008-0008129333 2   6.816718639403371 ...
851 ... BLOG06-20051206-040-0016432377 3   6.781155709916944 ...
851 ... BLOG06-20051206-056-0012738182 4   6.676189663420367 ...
                    ...
```

Roughly speaking, the main idea here is to re-rank these baselines according to the contents of our lexicons. The obtained benefit will be quantified by means of the standard tool *trec_eval*, which returns the values of the *MAP*, the *P@10* and the *R-Prec* [7] for any given baseline and *Qrels* file.

4.2 The IR Framework and Its Settings

Terrier [28] is the open-source framework for the rapid development of large-scale IR applications adopted for our experimentation. Terrier provides indexing and retrieval functionalities, and includes efficient and effective state-of-the-art retrieval models, for both document ranking and query expansion. From this point of view, Terrier is the only open-source platform natively supporting the DFR framework introduced in [1] and, as a consequence, is particularly suitable for our needs.

As concerns the setting of Terrier, according to the experimentation goals fixed in section 4, no stop-word list has been applied in the indexing process. Terms have been stemmed by the weak-Porter Stemming algorithm, which unlike the original version of the algorithm [35], does note drastically reduce terms to their radix form, making it possible to maintain different expressive forms with the same radix. For example "*like*" and "*likeable*" are kept as they are in the index.

5 Experimentation Description and Results

The following two subsections report on our experimentation activity, whose main goals and motivations have been illustrated in section 4.

As already stated, the main idea underlying our approach is that in order for a term to belong to the lexicon, it should appear in a suitable number of blog posts (that is, its document frequency should be greater than some threshold value), and that such threshold can be kept high enough to keep the size of the lexicon low, without reducing its effectiveness. In our experimentation, the following thresholds: $1, 100, 500, 1.000, 3.000, 5.000, 8.000, 10.000, 15.000, 20.000$, denoted $k_1, k_2, \ldots, k_{10}$ in what follows, have been considered.

To assess the stability of our lexicon an n-fold cross-validation approach [16, 9, 12], is followed. In n-fold cross-validation, the original collection is partitioned into n subsamples $s_1, \ldots, s_n$. The validation is performed in n phases (folds), where at phase i subsample s_i is retained as testing set, while the union $\cup_{j \neq i} s_j$ of the remaining $n-1$ subsamples is used as training set. The n results from the folds are then averaged (or otherwise combined) to produce a single estimation.

We performed a 5-fold cross-validation: let us denote by Tr_i and Ts_i, $i = 1, 2, \ldots, 5$, our training and testing sets, respectively. All training and testing sets have been generated from the set of 150 topics provided by NIST and described in section 4: such set of topics has been randomly partitioned into 5 equally sized subsets, named $Topics_i$, $i = 1, 2, \ldots, 5$ in what follows.

Next, each Ts_i has been obtained as the set of blog posts that have been classified as relevant with respect to the topics in $Topics_i$ and, consequently, Tr_i as its complementary to the set of all relevant blog posts. More formally: let $\langle top_b, id_b, judg_b \rangle$ denote any entry in the $Qrels$, where id_b is the identifier of the blog post b and $judg_b$ is the relevance/sentiment judgment expressed on b with respect to the topic top_b, and let

$$Qrels_i \equiv \{\langle top_b, id_b, judg_b \rangle \,|\, \langle top_b, id_b, judg_b \rangle \in Qrels, top_b \in Topics_i\},$$

and

$$Blogs_i \equiv \{b | \exists \langle top_b, id_b, judg_b \rangle \in Qrels_i, judg_b \geq 1\},$$

then

$$Ts_i \equiv Blogs_i$$

and

$$Tr_i = \bigcup_{1 \le j \le 5, j \ne i} Blogs_j,\ i = 1,2,\ldots,5$$

Once our training and testing sets have been generated, we also checked whether they approximately contain the same ratio, of opinionated posts, repeating the folds identification process as long as they do not substantially differ in percentage of opinionated documents.

Starting from each training set Tr_i, and for each tested value of the document frequency k_j, an opinion lexicon $Lex_{i,j}$ has been generated of all the terms t appearing in at least k_j blog posts of Tr_i. More formally:

$$Lex_{i,j} \equiv \left\{ t | DF_{t,Tr_i} \ge k_j \right\},$$

where DF_{t,Tr_i} is the document frequency of t inside Tr_i, with $i = 1,2,\ldots,5$ and $j = 1,2,\ldots,10$.

WordNet [13], an electronic lexical database for the English language created with the purpose of supporting automatic text analysis, is then accessed to recognize and discard all non-english terms from each $Lex_{i,j}$.

To complete the learning process, all terms in each $Lex_{i,j}$ have been weighted by a DFR-based query expansion model [5, 6]. As already stated, DFR models measure the divergence of a term distribution in the whole collection, Tr_i, from its distribution in a pseudo-relevance set, the subset of Tr_i made by the blog posts classified as opinionated in the *Qrels*: in our case, the higher is the divergence, the more the term is opinionated. Both the *KL* and *Bo1* query expansion, parameter free, models have been applied with this aim, as already stated in section 4.

For what concerns the testing process, as already stated, evaluations have be carried out by means of the *trec_eval* tool. In more detail, we considered *MAP* as the main measure according to which comparisons are carried out; however, for the sake of completeness, we also report *P@10* and *R-Prec* [7] values.

Table 1 *MAP* of reference in evaluating the effectiveness and the stability of our approach

BL	*MAP*	*P@10*	*R-Prec*
BL_1	0.2639	0.4753	0.3189
BL_2	0.2657	0.5287	0.3189
BL_5	0.3147	0.5307	0.3709
BL_3	0.3201	0.5387	0.3647
BL_4	0.3543	0.5580	0.3979

Then, reference values for the evaluations have been fixed, obtaining them from the baselines provided by NIST (see section 4) and denoting them as $BL_1, BL_2, \ldots, BL_5$. These values have been determined as follows: let $BL_{i,j}$ denote the set of entries of BL_i relevant with respect to topics in Ts_j. More formally: let $\langle top_b, Id_b, rank_b, score_b \rangle$ denote any entry of BL_i, where Id_b is the identifier assigned to the blog post b, commenting upon the topic top_b, in measure quantified by $score_b$, determining the rank $rank_b$ of b in BL_i, and let

$$BL_{i,j} \equiv \left\{ \langle top_b, Id_b, rank_b, score_b \rangle \mid \langle top_b, Id_b, rank_b, score_b \rangle \in BL_i, top_b \in Ts_j \right\},$$

where $i, j = 1, 2, \ldots, 5$.

The value of reference MAP_i, for each baseline BL_i, are shown in table 1. Since we argue that the higher is the MAP of a baseline, the lower is the benefit achievable by our lexicons, in order to make comparisons easier, baseline entries are arranged in order of increasing values of MAP_i.

Notice that, from these baselines, we already obtain rankings of topic relevance. As a consequence, we may skip the fist step of the procedure introduced in section 3 and we may accomplish only the last two steps, comparing then the evaluation measures of the obtained re-rankings with the ones of the baselines.

Once the reference values have been fixed, we evaluate the effectiveness of the approach for all generated lexicons. More precisely, for each baseline BL_i ($i = 1, 2, \ldots, 5$) and for each testing set Ts_j ($j = 1, 2, \ldots, 5$), $BL_{i,j}$ have been boosted through $Lex_{j,l}$, for each value of the document frequency k_l ($l = 1, 2, \ldots, 10$): let $R\text{-}BL_{i,j,l}$ denote the resulting new run.

The effectiveness of this process, as already stated, is evaluated by means of *trec_eval*: let $MAP_{i,j,l}$ denote the value of the MAP of $R\text{-}BL_{i,j,l}$ with respect to $Qrels_j$. Then, for each BL_i ($i = 1, 2, \ldots, 5$) and for each k_l ($l = 1, 2 \ldots, 10$), the mean $\overline{MAP}_{i,l} = \sum_{j=1}^{5} \frac{1}{5} MAP_{i,j,l}$, has been computed.

As already stated, we first of all focus on the *KL-Dfreel* combination of DFR models. The results of our experiments are reported in table 2, which shows, for each baseline BL_i ($1, 2, \ldots, 5$) and for each considered document frequency k_j ($j = 1, 2, \ldots, 10$), the value of $\overline{MAP}_{i,j}$ and the percentage variation of $\overline{MAP}_{i,j}$ with respect to MAP_i. The table also shows the average size $\overline{Size}_j$ of the lexicons $Lex_{1,j}, Lex_{2,j}, \ldots, Lex_{5,j}$. Maximum values of $\overline{MAP}_{i,j}$ are highlighted in boldface.

We also compute the standard deviation of $MAP_{i,j,l}$ with respect to Ts_j. Table 3 shows these values, $\sigma_{i,j}$, for each baseline BL_i and each value of the document frequency k_j ($i = 1, 2, \ldots, 5$, $j = 1, 2, \ldots, 10$).

Let us briefly discuss the results reported in tables 2 and 3. First of all, it is worth to notice that $\Delta M_{i,j}^{\%} > 0$, independently from the training set and the document frequency threshold. Thus, in the average, our approach introduces a relevant benefit with respect to IR systems not supporting any opinion-finding feature. Furthermore, as k_j grows, $\Delta M_{i,j}^{\%}$ increases up to a maximum, and then starts decreasing. This reflects the intuition that too crowd lexicons, as also too poor ones, are not able to effectively capture the content of opinion of a blog post. However, the effectiveness of the derived lexicons is not greatly influenced by their size: in fact, for all baselines

Table 2 Effectiveness of the *KL-Dfree1* combination of DFR models in the re-ranking process

KL-DFree1 Cross validation - *MAP*											
		BL_1		BL_2		BL_5		BL_3		BL_4	
k_j	$\overline{Size_j}$	$\overline{MAP_{1,j}}$	$\Delta M^{\%}_{1,j}$	$\overline{MAP_{2,j}}$	$\Delta M^{\%}_{2,j}$	$\overline{MAP_{5,j}}$	$\Delta M^{\%}_{5,j}$	$\overline{MAP_{3,j}}$	$\Delta M^{\%}_{3,j}$	$\overline{MAP_{4,j}}$	$\Delta M^{\%}_{4,j}$
1	8627.6	0.3024	14.60%	0.2737	3.03%	0.3378	7.33%	0.3448	7.71%	0.3731	5.32%
100	4711.8	0.3024	14.60%	0.2737	3.03%	0.3378	7.33%	0.3448	7.71%	0.3731	5.31%
500	2177.4	0.3024	14.60%	0.2738	3.06%	0.3377	7.31%	0.3448	7.72%	0.3732	5.33%
1000	1452.8	0.3024	14.60%	0.2740	3.13%	0.3377	7.31%	**0.3449**	**7.74%**	0.3732	5.33%
3000	606.2	0.3025	14.63%	0.2749	3.46%	0.3380	7.41%	0.3448	7.70%	0.3735	5.41%
5000	364.4	**0.3027**	**14.70%**	0.2756	3.74%	0.3383	7.51%	0.3448	7.72%	0.3738	5.52%
8000	223.8	0.3023	14.54%	0.2767	4.15%	**0.3384**	**7.53%**	0.3448	7.72%	**0.3745**	**5.71%**
10000	147.6	0.3020	14.42%	0.2772	4.33%	0.3381	7.45%	0.3444	7.59%	0.3743	5.66%
15000	65.2	0.2992	13.39%	**0.2774**	**4.42%**	0.3345	6.28%	0.3434	7.27%	0.3734	5.39%
20000	18.4	0.2936	11.26%	0.2751	3.55%	0.3273	4.00%	0.3402	6.27%	0.3680	3.86%

KL-DFree1 Cross validation - *P@10*										
	BL_1		BL_2		BL_5		BL_3		BL_4	
k_j	$\overline{P@10_{1,j}}$	$\Delta P^{\%}_{1,j}$	$\overline{P@10_{2,j}}$	$\Delta P^{\%}_{2,j}$	$\overline{P@10_{5,j}}$	$\Delta P^{\%}_{5,j}$	$\overline{P@10_{3,j}}$	$\Delta P^{\%}_{3,j}$	$\overline{P@10_{4,j}}$	$\Delta P^{\%}_{4,j}$
1	0.5313	11.79%	0.5387	1.89%	0.5760	8.54%	0.5787	7.42%	0.5867	5.14%
100	0.5313	11.79%	0.5387	1.89%	0.5760	8.54%	0.5787	7.42%	0.5867	5.14%
500	0.5313	11.79%	0.5393	2.01%	0.5760	8.54%	0.5787	7.42%	0.5873	5.26%
1000	0.5300	11.51%	0.5393	2.01%	0.5760	8.54%	0.5793	7.54%	0.5873	5.26%
3000	0.5313	11.79%	0.5373	1.63%	0.5773	8.79%	0.5793	7.54%	0.5867	5.14%
5000	0.5320	11.93%	0.5373	1.63%	0.5767	8.66%	0.5787	7.42%	0.5887	5.49%
8000	0.5320	11.93%	0.5380	1.76%	0.5780	8.91%	0.5800	7.67%	0.5874	5.26%
10000	0.5307	11.65%	0.5393	2.01%	0.5760	8.54%	0.5787	7.42%	0.5873	5.26%
15000	0.5273	10.95%	0.5393	2.01%	0.5700	7.41%	0.5800	7.67%	0.5873	5.26%
20000	0.5227	9.96%	0.5373	1.63%	0.5673	6.90%	0.5720	6.19%	0.5860	5.02%

KL-DFree1 Cross validation - *R-Prec*										
	BL_1		BL_2		BL_5		BL_3		BL_4	
k_j	$\overline{RPrec_{1,j}}$	$\Delta RP^{\%}_{1,j}$	$\overline{RPrec_{2,j}}$	$\Delta RP^{\%}_{2,j}$	$\overline{RPrec_{5,j}}$	$\Delta RP^{\%}_{5,j}$	$\overline{RPrec_{3,j}}$	$\Delta RP^{\%}_{3,j}$	$\overline{RPrec_{4,j}}$	$\Delta RP^{\%}_{4,j}$
1	0.3568	11.90%	0.3263	2.33%	0.3951	6.53%	0.3880	6.38%	0.4145	4.17%
100	0.3568	11.90%	0.3264	2.35%	0.3951	6.53%	0.3880	6.38%	0.4145	4.16%
500	0.3568	11.90%	0.3264	2.34%	0.3952	6.56%	0.3880	6.38%	0.4149	4.28%
1000	0.3568	11.90%	0.3264	2.36%	0.3952	6.56%	0.3881	6.41%	0.4150	4.29%
3000	0.3565	11.80%	0.3271	2.58%	0.3952	6.55%	0.3876	6.29%	0.4150	4.29%
5000	0.3571	11.98%	0.3277	2.75%	0.3953	6.58%	0.3882	6.45%	0.4156	4.44%
8000	0.3555	11.48%	0.3281	2.88%	0.3955	6.64%	0.3891	6.70%	0.4148	4.26%
10000	0.3548	11.27%	0.3282	2.91%	0.3953	6.59%	0.3889	6.65%	0.4144	4.16%
15000	0.3531	10.71%	0.3288	3.12%	0.3910	5.43%	0.3880	6.39%	0.4133	3.88%
20000	0.3461	8.53%	0.3266	2.43%	0.3866	4.23%	0.3827	4.95%	0.4072	2.34%

the difference between the maximum and the minimum achieved benefit is quite small. Hence, it is possible to keep the document frequency threshold high enough to maintain the size of the lexicon tractable, without losing much of its effectiveness.

When considering standard deviation values $\sigma_{i,j}$ for the MAP, we may notice that they are always small, thus confirming the substantial independence of our method

Table 3 Dependence of $MAP_{i,j,l}$ from the particular testing set for the *KL-Dfree1* combination of DFR models

KL-DFree1					
	BL_1	BL_2	BL_5	BL_3	BL_4
k_j	$\sigma_{1,j}$	$\sigma_{2,j}$	$\sigma_{5,j}$	$\sigma_{3,j}$	$\sigma_{4,j}$
1	4.50%	4.25%	5.03%	4.86%	4.60%
100	4.50%	4.25%	5.03%	4.86%	4.60%
500	4.49%	4.25%	5.04%	4.86%	4.60%
1000	4.48%	4.23%	5.04%	4.85%	4.60%
3000	4.47%	4.25%	5.04%	4.85%	4.61%
5000	4.47%	4.25%	5.02%	4.83%	4.60%
8000	4.43%	4.31%	4.99%	4.79%	4.62%
10000	4.41%	4.31%	4.94%	4.77%	4.62%
15000	4.34%	4.33%	4.96%	4.71%	4.58%
20000	4.44%	4.28%	4.58%	4.66%	4.43%

from the training sets. Taking a look to the maximum values of $\Delta M^{\%}_{i,j}$, it is also clear that our assumption that the higher is the *MAP* of a baseline, the lower is, reasonably, the benefit achieved by our lexicons, is confirmed, with the only exception of BL_2. The reason why the intuition is not verified for this baseline is not clear at the moment, and it may require a deeper investigation.

Concerning *P@10* and *R-Prec* values, also in this case we always obtain an increment on the original baseline values. Furthermore, the difference between the best and the worst improvement is quite small. Differently from the *MAP* case, and in particular for *P@10*, the values trend is not so regular. Finally, the *MAP* values always overcome the increases obtained in the other two measure. All these observations allow us to assert that our method does not affect only early precision, but also performs a substantial improvement in the whole ranking.

Despite of the assumed suitability of *KL* to approximate the opinion entropy function, we also repeated our evaluation applying the *Bo1-DFree1* combination of DFR models. *Bo1* is one of the most effective DFR Query Expansion models, based on the Bose-Einstein distribution: compared with *KL*, it tends to identify a more broad set of relevant terms for a given document. This implies that *Bo1* identifies more candidate terms with respect to *KL* and the generated lexicons have a greater number of opinionated terms.

Table 4 shows the results of our experimentation. The most important observation here is that the achieved benefit is, for all the baselines and for all the document frequency thresholds, outperformed by the benefit introduced by the *Bo1-DFree1* solution, with the only exception of the baseline BL_5 and $k_j > 15000$. Furthermore, it is worth to notice that, in some cases, the re-ranked baseline is outperformed by IR systems not supporting any opinion-finding feature (negative values of $\Delta M^{\%}_{i,j}$). These observations enforce the effectiveness of the *KL* choice. In fact, what we really need is not just a query expansion model, but also a good approximation of the opinion entropy function.

Table 4 Effectiveness of the *Bo1-Dfree1* combination of DFR models in the re-ranking process

Bo1-DFree1 Cross validation											
		BL_1		BL_2		BL_5		BL_3		BL_4	
k_j	$Size_j$	$MAP_{1,j}$	$\Delta M^{\%}_{1,j}$	$MAP_{2,j}$	$\Delta M^{\%}_{2,j}$	$MAP_{5,j}$	$\Delta M^{\%}_{5,j}$	$MAP_{3,j}$	$\Delta M^{\%}_{3,j}$	$MAP_{4,j}$	$\Delta M^{\%}_{4,j}$
1	14281.5	0.2701	2.36%	0.2536	-3.92%	0.3070	-2.45%	0.3160	-1.29%	0.3505	-1.08%
100	8705.25	0.2698	2.24%	0.2530	-4.79%	0.3178	0.97%	0.3179	-0.70%	0.3475	-1.93%
500	4030.75	0.2698	2.23%	0.2538	-4.49%	0.3182	1.11%	0.3179	-0.68%	0.3479	-1.80%
1000	2640.8	0.2873	8.87%	0.2584	-2.74%	0.3266	3.79%	0.3338	4.29%	0.3612	1.94%
3000	1041.2	0.2889	9.47%	0.2631	-0.96%	0.3289	4.52%	0.3346	4.52%	0.3641	2.77%
5000	591.2	0.2902	9.97%	0.2678	0.78%	0.3321	5.54%	0.3353	4.74%	0.3676	3.75%
8000	281	0.2923	10.76%	0.2728	2.67%	0.3344	6.26%	0.3375	5.43%	0.3705	4.56%
10000	199.4	0.2934	11.18%	0.2743	3.24%	0.3355	6.60%	0.3381	5.62%	0.3713	4.80%
15000	**75.2**	**0.2941**	**11.44%**	**0.2756**	**3.71%**	**0.3365**	**6.91%**	**0.3386**	**5.77%**	**0.3716**	**4.88%**
20000	19.6	0.2881	9.19%	0.2729	2.71%	0.3296	4.75%	0.3357	4.86%	0.3661	3.34%

Table 5 Effectiveness of the *KL-DPH* combination of DFR models in the re-ranking process

KL-DPH Cross validation											
		BL_1		BL_2		BL_5		BL_3		BL_4	
k_j	$Size_j$	$MAP_{1,j}$	$\Delta M^{\%}_{1,j}$	$MAP_{2,j}$	$\Delta M^{\%}_{2,j}$	$MAP_{5,j}$	$\Delta M^{\%}_{5,j}$	$MAP_{3,j}$	$\Delta M^{\%}_{3,j}$	$MAP_{4,j}$	$\Delta M^{\%}_{4,j}$
1	8627.6	0.3034	14.98%	0.2768	4.17%	0.3324	5.62%	0.3472	8.48%	0.3549	0.16%
100	4711.8	0.3035	15.00%	0.2768	4.17%	0.3324	5.62%	0.3473	8.48%	0.3549	0.16%
500	2177.4	0.3035	15.01%	0.2771	4.28%	0.3324	5.62%	0.3473	8.48%	0.3549	0.16%
1000	1452.8	0.3035	15.01%	0.2773	4.36%	0.3324	5.62%	0.3473	8.49%	0.3549	0.18%
3000	606.2	0.3037	15.09%	0.2783	4.76%	0.3325	5.64%	0.3474	8.52%	0.3552	0.24%
5000	364.4	**0.3040**	**15.20%**	0.2783	4.76%	**0.3378**	**7.35%**	0.3478	8.66%	**0.3724**	**5.10%**
8000	223.8	0.3036	15.03%	0.2792	5.07%	0.3376	7.28%	0.3481	8.73%	0.3721	5.02%
10000	147.6	0.3033	14.94%	0.2800	5.37%	0.3368	7.02%	**0.3481**	**8.75%**	0.3717	4.91%
15000	65.2	0.3009	14.01%	**0.2802**	**5.44%**	0.3292	4.61%	0.3481	8.75%	0.3526	-0.47%
20000	18.4	0.2857	8.27%	0.2705	1.79%	0.3171	0.76%	0.3423	6.94%	0.3419	-3.51%

Table 6 Final validation of the effectiveness of the *KL-Dfree1* combination of DFR models in the re-ranking process

KL-DFree1 Final validation											
		BL_1		BL_2		BL_5		BL_3		BL_4	
k_j	$Size_j$	$MAP_{1,j}$	$\Delta M^{\%}_{1,j}$	$MAP_{2,j}$	$\Delta M^{\%}_{2,j}$	$MAP_{3,j}$	$\Delta M^{\%}_{3,j}$	$MAP_{4,j}$	$\Delta M^{\%}_{4,j}$	$MAP_{5,j}$	$\Delta M^{\%}_{5,j}$
1	8529	0.3010	14.06%	0.2742	3.20%	0.3332	5.88%	0.3452	7.84%	0.3731	5.31%
100	5125	0.3010	14.06%	0.2742	3.20%	0.3332	5.88%	**0.3452**	**7.84%**	0.3731	5.31%
500	2455	0.3010	14.06%	0.2742	3.20%	0.3332	5.88%	0.3451	7.81%	0.3731	5.31%
1000	1668	0.3010	14.06%	0.2743	3.24%	0.3332	5.88%	0.3451	7.81%	0.3730	5.28%
3000	750	0.3009	14.02%	0.2748	3.42%	0.3333	5.91%	0.3450	7.78%	0.3732	5.33%
5000	463	0.3011	14.10%	0.2758	3.80%	0.3336	6.01%	0.3451	7.81%	0.3741	5.59%
8000	271	**0.3011**	**14.10%**	0.2764	4.03%	0.3336	6.01%	0.3450	7.78%	0.3741	5.59%
10000	208	0.3010	14.06%	0.2766	4.10%	**0.3337**	**6.04%**	0.3451	7.81%	**0.3742**	**5.62%**
15000	111	0.2997	13.57%	0.2773	4.37%	0.3331	5.85%	0.3441	7.50%	0.3734	5.39%
20000	57	0.2973	12.66%	**0.2774**	**4.40%**	0.3313	5.27%	0.3429	7.12%	0.3723	5.08%

Table 7 Final validation of the effectiveness of the *KL-DPH* combination of DFR models in the re-ranking process

KL-DPH Final validation											
		BL_1		BL_2		BL_5		BL_3		BL_4	
k_j	$Size_j$	$MAP_{1,j}$	$\Delta M^{\%}_{1,j}$	$MAP_{2,j}$	$\Delta M^{\%}_{2,j}$	$MAP_{3,j}$	$\Delta M^{\%}_{3,j}$	$MAP_{4,j}$	$\Delta M^{\%}_{4,j}$	$MAP_{5,j}$	$\Delta M^{\%}_{5,j}$
1	8529	0.3046	15.42%	0.2769	4.22%	0.3399	8.01%	0.3480	8.72%	0.3728	5.22%
100	5125	0.3046	15.42%	0.2769	4.22%	0.3399	8.01%	0.3480	8.72%	0.3728	5.22%
500	2455	0.3046	15.42%	0.2770	4.25%	0.3398	7.98%	0.3479	8.68%	0.3728	5.22%
1000	1668	0.3046	15.42%	0.2771	4.29%	0.3398	7.98%	0.3479	8.68%	0.3729	5.25%
3000	750	0.3047	15.46%	0.2777	4.52%	0.3399	8.01%	0.3479	8.68%	0.3727	5.19%
5000	463	0.3051	15.61%	0.2784	4.78%	0.3401	8.07%	0.3480	8.72%	0.3731	5.31%
8000	271	0.3052	15.65%	0.2794	5.16%	0.3402	8.10%	0.3484	8.84%	0.3729	5.25%
10000	208	**0.3055**	**15.76%**	0.2821	6.17%	**0.3415**	**8.52%**	**0.3490**	**9.03%**	**0.3749**	**5.81%**
15000	111	0.3036	15.04%	**0.2828**	**6.44%**	0.3389	7.69%	0.3485	8.87%	0.3725	5.14%
20000	57	0.2971	12.58%	0.2751	3.54%	0.3255	3.43%	0.3464	8.22%	0.3602	1.67%

Since *KL* is confirmed as an effective choice, we once again repeat our experiments applying the *KL-DPH* combination of DFR models. The results are shown by table 5: it is clear from such results that *DPH* and *DFreeI* have similar performance in terms of the introduced benefits and of the size of the lexicons required to achieve these improvements. We may also notice that *DPH* seems to outperform *DFreeI* when the baseline is difficult. Anyway, a deeper investigation would be required to give this observation a statistical relevance.

To conclude the cross validation process, a final experimentation has been conducted by extracting the lexicons from the whole set of opinionated blog posts and testing their effectiveness on BL_i, $i = 1, 2, \ldots 5$. Both the *KL-DFreeI* and the *KL-DPH* combinations of DFR models have been considered. Tables 6 and 7 show the results of both experimentations.

The comparison of table 2 with 6 and of table 5 with 7, substantially confirms the results of our cross validation process.

Let us now compare the results we achieved at the last TREC Blog Track to the ones obtained by other participants. We report in table 8 the median, worst and best values of *MAP* for all the participants with respect to baseline 4, which has the highest opinion finding MAP. Those values are related to the new 50 topics of TREC 2008.

Table 8 Median, best and worst MAP values and their percentage variation, for all participants to TREC 2008 on baseline 4, which is the best one with respect to opinion MAP

	MAP	$\Delta M^{\%}_4$
median	0.3964	3.72%
best	0.4189	9.60%
worst	0.2341	-38.75%
our	0.4006	4.81%

It is worth to notice that the median increase of BL_4 is small, confirming that improving a good baseline is difficult. In addition, our approach shows a substantial effectiveness when compared to the results obtained by TREC participants.

Finally, let us compare our lexicon based approach with the ones proposed by He *et al.* in [14]: in these papers different approaches, some of them also relying on proximity evaluation, are introduced and studied. We limit ourselves to consider only the simpler model presented, where proximity is not exploited.

Both models are based on both a DFR approach to the automatic construction of an opinion lexicon and a re-ranking strategy. However, the modalities of lexicon construction are quite different, as well as the re-ranking function itself. For more details see [14, 38]. This comparison allows a final evaluation of the effectiveness obtainable by lexicon based approaches. On table 9, we consider the *MAP* values obtained on the 5 TREC baselines for the topics of the last two TREC editions, listing the ones resulting by our approach in the final validation with *KL-DPH* models combination, those obtained in [14] approach as reported in [38] (referred in the table as HMHO), and the TREC median. Also in this case we sorted the baselines, according to the increasing order of *MAP* values.

Table 9 Comparison between our lexicon based approach to opinion mining, the one proposed in [14] and the median values of TREC 2008 participants

topic 2007	BL1	BL2	BL3	BL4	BL5	
baseline	0.2758	0.3034	0.3489	0.3784	0.3805	$\overline{\Delta M}\%$
HMHO	0.2988	0.3020	0.3561	0.3885	0.3839	2.70%
KL-DPH	0.3245	0.3320	0.3858	0.4090	0.4149	10.84%
TREC median	0.3077	0.3298	0.4128	0.3950	0.3709	9.12%

topic 2008	BL2	BL5	BL1	BL3	BL4	
baseline	0.2639	0.2988	0.3239	0.3564	0.3822	$\overline{\Delta M}\%$
HMHO	0.2621	0.3008	0.3512	0.3669	0.3964	3.02%
KL-DPH	0.2776	0.3215	0.3681	0.3848	0.3997	7.78%
TREC median	0.2705	0.3010	0.3493	0.3705	0.3848	0.76%

Notice that our lexicon based approach always overcomes the TREC median, with only one exception. In general, also the approach in [14] exceeds the TREC median, showing its effectiveness. While *KL-DPH* seems to be more effective, the evaluation is not definitive, due to the use of different training sets. Besides, the values reported for [14], are related to the simpler method they propose, which is the most similar to our model. The MAP improvements obtained on each baseline by these two approaches maintain in fact an approximately constant gap, showing a similar behavior of the models.

5.1 Lexicons Composition

Finally, we investigate the composition of the generated lexicons, mainly focusing on the presence of stop-words. We will compare the content of each $Lex_{i,j}$, generated in the cross validation process, with the one of the stop-word list provided by Terrier. To accomplish this task, stop words have been stemmed by the same stemming algorithm adopted to pre-processing the collection of blog posts, i.e. the Weak-Porter Stemmer. Table 10 shows the results of this investigation.

Table 10 Overlap between Lex_j, for each level k_j, and a standard stop words list of english terms provided by Terrier [28]

	KL			*Bo1*		
k_j	$\overline{Size_j}$	$SWL \cap Lex_j$	%	$\overline{Size_j}$	$SWL \cap Lex_j$	%
1	8627.6	229.8	2.66%	14281.5	295.5	2.07%
100	4711.8	221	4.69%	8705.25	290.25	3.33%
500	2177.4	209	9.60%	4030.75	266	6.60%
1000	1452.8	200.2	13.78%	2640.8	253.4	9.60%
3000	606.2	166.6	27.48%	1041.2	205.2	19.71%
5000	364.4	137.6	37.76%	591.2	165	27.91%
8000	223.8	115	51.39%	281	122.6	43.63%
10000	147.6	92.8	62.87%	199.4	100	50.15%
15000	65.2	52.4	80.37%	75.2	54.4	72.34%
20000	18.4	18	97.83%	19.6	18	91.84%

The results are quite surprising, since stop-words are usually considered either noisy terms (i.e., articles, prepositions) or so common that their presence should not really characterize a document at all. For the opinion mining task instead, they turn out to be useful. It is quite impressive, in fact, that the best performing lexicons, for both *KL* and *Bo1*, are the ones obtained for a document frequency higher than 8000, i.e. the ones showing a predominant presence of stop-words (at least half of the lexicon in most of the cases). This means that some stop words are opinion bearing, i.e. are used so much differently in opinionated documents that in objective ones, to be statistically relevant in opinion detection.

6 Polarity Detection

In the previous section, we show that our lexicon based method is effective in identifying and weighting documents according to their opinion content. One can argue whether the same approach gives comparable benefits when applied to polarity detection, i.e. to the identification and the weighting of blog posts on the basis of the polarity (positive or negative) of the opinion they express. Unfortunately, this does not seem to be the case, as we experienced in our participation to TREC 2008 blog track [31]. In what follows we briefly report on this participation and close the section with some alternative approaches we started to follow trying to improve the poor results we achieved.

We assumed that if we can statistically identify terms bearing an opinion content, we can, in the same way, identify terms that are used to express positive or negative opinions. Consequently, we tried to predict the polarity orientation of a blog post just by observing the distribution of its terms. With this aim, analogously to what described in section 4, we defined and approximated an *average positive* and *negative entropy function*.

By these approximations we built two *polarized opinion lexicon*, for positive and negative terms respectively. We started by the same assumptions followed for the opinion lexicon, exploiting the divergence between the distribution of terms in in the set of blog posts expressing a positive (resp., negative) opinion on some topics and the set of all the opinionated blog posts for the same topics.

In this process of construction, contrary to the opinion retrieval case, we assumed that stop-words only introduce noise when dealing with polarity and consequently discarded them from each blog post in the collection. For example, while stop-words like "*not*" or "*I*" seems to bear an opinion content, they does not seem to contribute to the polarity of the expressed opinion.

As far as the polarity recognition process is concerned, we quantified the benefit introduce with respect to IR systems which do not support any polarity-finding feature. In the approach we considered, the polarity rank of a blog post is determined starting from the weights assigned to polarity bearing terms in the two lexicons: the post has assigned a positive and a negative score of polarity, while its final score, the one determining its position in the ranking, is obtained as a difference of the two. If this final score is positive, the document is classified as expressing a positive opinion; a negative opinion, otherwise. Finally if the score is close to zero, the document has been considered to be *mixed*.

Unfortunately, this approach gave poor results but, even worse, also most of the participants to the TREC blog track experienced the same unsatisfactory performances (most of the submitted runs resulted to be worse than the starting baseline). This seems to affirm that polarity detection is really a challenging task to deal with, as shown by table 11, borrowed from [31], that provides the average best, median and worst MAP measures for each of the 150 topics considered in this paper, across all 2006-2008 years, for all submitted baselines and runs.

Table 11 Best, median, worst of baseline and polarity runs over all 150 topics of Blog Track participants [31]

	Baselines runs			Polarity runs		
	MAP_{pos}	MAP_{neg}	MAP_{mix}	MAP_{pos}	MAP_{neg}	MAP_{mix}
median	0.1143	0.0751	0.0964	0.1143	0.0648	0.0933
best	0.2367	0.1676	0.1854	0.4006	0.4292	0.3693
worst	0.0144	0.0055	0.0114	0.0028	0.0011	0.0031

Since the lexicon based approach we considered is not successful in polarity detection, it seems necessary to change perspective of analysis: in particular, relaxing the bag of words hypothesis appears a necessary step towards the definition of more effective techniques. In particular, passage retrieval techniques [15, 21, 36] could be applied to try to maintain the relationship among words in the whole document or in some of its portion, such as chapters, paragraphs, sentences, fixed or variable size windows of text. Using passage retrieval in polarity retrieval has been proved useful by several participants to the TREC [29, 24, 31], even if in combination with other techniques [42, 19].

A different approach is the one based on the concept of *cover*, i.e. a sequence of terms that begins and ends with a term of the query [27, 19]. In particular in [27] it has been proved that, if the weights are assigned to the covers by a *length normalized scoring function*, then the best passage, i.e. the passage that maximizes the scoring function, is a cover.

In general, however, the efficiency and effectiveness of such type of techniques for polarity retrieval have still to be more completely characterized.

7 Conclusions

The main aim of this work is to prove the effectiveness of the proposed model and to investigate its stability. In particular, since this approach relies on a training phase, it is relevant to verify that it maintains its effectiveness as the training set changes. For such a reason we performed a 5-fold cross-validation, evaluating the appropriateness of the model and attesting its stability. The overall performance is quite high, improving the baseline *MAP* by more than 10% in some cases.

Furthermore, the model maintains its effectiveness on lexicons of different size, showing that it is possible to obtain good results in opinion retrieval even by using small dictionaries.

The comparison between *KL* and *Bo1*, asserts that the quality of the opinion retrieval method presented here is greatly dependent from the underlying dictionary. It is also clear that *KL* obtains better performance than *Bo1*, also generally returning smaller lexicons. This evidence reinforce the validity of our theoretical model in identifying opinion bearing terms. *KL* in fact, differently from *Bo1*, is a measure of terms distribution divergence that is used as approximation of the opinion entropy function.

It is also worth to notice that the analysis of lexicons has shown how a lot of terms identified as opinion bearing are characterized by a low information content. This suggests that also some common stop words could be useful in the identification of opinion expressions. A deeper analysis should be performed to identify more precisely these terms and to investigate their relationship with well known opinion bearing terms like verbs or adjectives. This follows the assumption that there are terms that links concepts in the sentences. For examples in [10] is reported a study on the role of adjectives and verbs affirming that, since *verbs* link nouns, also *verbs* possibly testify presence of opinions.

Finally, the results obtained by applying the *DPH* weighting model to assign opinion scores to documents shows instead how it is possible to improve the performance of the system. A further investigation on the best weighting model should be accomplished.

For what regards polarity detection, it seems that an approach derived from the one used for opinion retrieval would not be successful. A bag of words approach seems to be not sufficient to effectively tackle the problem, as also confirmed by TREC Blog tracks results [29, 24, 31]. For polarity detection, in fact, we must focus not only on single terms appearing in a document, but also on their relationships. For example it has been shown that in subjective documents the presence of *adverbs* surrounding adjectives influences the strength of opinion expressed [8]. Moreover, in English a noun follows an adjective with probability 0.57 [39], suggesting that either proximity analysis [37] or passage retrieval [19] could be suitably applied to the problem.

References

1. Amati, G.: Probability Models for Information Retrieval based on Divergence from Randomness. PhD thesis, Department of Computing Science, University of Glasgow (2003)
2. Amati, G.: Frequentist and bayesian approach to information retrieval. In: Proc. of the 28th European Conference on IR Research (ECIR) (2006)
3. Amati, G., Ambrosi, E., Bianchi, M., Gaibisso, C., Gambosi, G.: Fub, iasi-cnr and university of tor vergata at trec 2007 blog track. In: Proc. of the 16th Text Retrieval Conference (TREC) (2007)
4. Amati, G., Ambrosi, E., Bianchi, M., Gaibisso, C., Gambosi, G.: Automatic construction of an opinion-term vocabulary for ad hoc retrieval. In: Proc. of the 30th European Conference on IR Research, ECIR (2008)
5. Amati, G., Carpineto, C., Romano, G.: Fub at trec-10 web track: a probabilistic framework for topic relevance term weighting. In: Proc. of the 16th Text Retrieval Conference, TREC (2001)
6. Amati, G., Van Rijsbergen, C.J.: Probabilistic models of information retrieval based on measuring divergence from randomness. ACM Transactions on Information Systems 20(4), 357–389 (2002)
7. Baeza-Yates, R., Ribeiro-Neto, B.: Modern Information Retrieval. Addison-Wesley, Reading (1999)
8. Benamara, F., Cesarano, C., Picariello, A., Reforgiato, D.: Sentiment analysis: Adjectives and adverbs are better than adjectives alone. In: Proc. of the 1st International Conference on Weblogs and Social Media, ICWSM (2007)
9. Chang, J.S., Luo, Y.F., Su, K.Y.: Gpsm: A generalized probabilistic semantic model for ambiguity resolution. In: Proc. of the 30th annual meeting on Association for Computational Linguistics (1992)
10. Chesley, P., Vincent, B., Xu, L., Srihari, R.: Using verbs and adjectives to automatically classify blog sentiment. In: AAAI Spring Symposium on Computational Approaches to Analysing Weblogs (2006)
11. Dave, K., Lawrence, S., Pennock, D.M.: Mining the peanut gallery: opinion extraction and semantic classification of product reviews. In: Proceedings of the 12th International Conference on World Wide Web, WWW (2003)
12. Devijver, P.A., Kittler, J.: Pattern Recognition: A Statistical Approach. Prentice-Hall, Englewood Cliffs (1982)
13. Fellbaum, C. (ed.): WordNet: An Electronic Lexical Database (Language, Speech, and Communication). The MIT Press, Cambridge (1998)
14. He, B., Macdonald, C., He, J., Ounis, I.: An effective statistical approach to blog post opinion retrieval. In: Proc. of the 17th ACM conference on Information and knowledge management, CIKM (2008)

15. Kaszkiel, M., Zobel, J.: Passage retrieval revisited. In: Proc. of the 20th Annual International ACM SIGIR Conference on Research and Development in Information Retrieval, SIGIR (1997)
16. Kohavi, R.: A study of cross-validation and bootstrap for accuracy estimation and model selection. In: Proc. of the 14th International Joint Conference on Artificial Intelligence, IJCAI (1995)
17. Kullback, S.: The kullback-leibler distance. The American Statistician 41, 340–341 (1987)
18. Kullback, S., Leibler, R.A.: On information and sufficiency. Annals of Mathematical Statistics 22, 49–86 (1951)
19. Lee, Y., Na, S., Kim, J., Nam, S., Jung, H., Lee, J.: Kle at trec 2008 blog track: Blog post and feed retrieval. In: Proc. of the 17th Text Retrieval Conference, TREC (2008)
20. Liu, B.: Web Data Mining: Exploring Hyperlinks, Contents, and Usage Data (Data-Centric Systems and Applications). Springer, Heidelberg (2007)
21. Liu, X., Croft, W.B.: Passage retrieval based on language models. In: Proc. of the 11th ACM Conference on Information and Knowledge Management (CIKM), pp. 375–382 (2002)
22. Macdonald, C., He, B., Plachouras, V., Ounis, I.: University of glasgow at trec 2005: Experiments in terabyte and enterprise tracks with terrier. In: Proc. of the 14th Text Retrieval Conference, TREC (2005)
23. Macdonald, C., Ounis, I.: The trec blogs06 collection: Creating and analysing a blog test collection. DCS Technical Report Series (2006)
24. Macdonald, C., Ounis, I., Soboroff, I.: Overview of the trec-2007 blog track. In: Proc. of the 16th Text Retrieval Conference, TREC (2007)
25. Manmatha, R., Rath, T., Feng, F.: Modeling score distributions for combining the outputs of search engines. In: Proc. of the 24th Annual International ACM SIGIR Conference on Research and Development in Information Retrieval, SIGIR (2001)
26. Manning, C.D., Raghavan, P., Schütze, H.: Introduction to Information Retrieval. Cambridge University Press, Cambridge (2008)
27. Na, S., Kang, I., Lee, Y., Lee, J.: Completely-arbitrary passage retrieval in language modeling approach. In: Proc. of the 4th Asia Infomation Retrieval Symposium, AIRS (2008)
28. Ounis, I., Amati, G., Plachouras, V., He, B., Macdonald, C., Lioma, C.: Terrier: A high performance and scalable information retrieval platform. In: Proc. of the ACM SIGIR'06 Workshop on Open Source Information Retrieval, OSIR (2006)
29. Ounis, I., de Rijke, M., Macdonald, C., Mishne, G.A., Soboroff, I.: Overview of the trec-2006 blog track. In: TREC 2006 Working Notes (2006)
30. Ounis, I., Macdonald, C., Soboroff, I.: On the trec blog track. In: Proc. of the 2nd International Conference on Weblogs and Social Media, ICWSM (2008)
31. Ounis, I., Macdonald, C., Soboroff, I.: Overview of the trec-2008 blog track. In: Proc. of the 17th Text Retrieval Conference, TREC (2008)
32. Pang, B., Lee, L.: Opinion mining and sentiment analysis. Foundations and Trends in Information Retrieval 2(1–2), 1–135 (2008)
33. Pang, B., Lee, L., Vaithyanathan, S.: Thumbs up? sentiment classification using machine learning techniques. In: Proc. of the ACL 2002 conference on Empirical Methods in Natural Language Processing, pp. 79–86 (2002)
34. Plachouras, P., He, B., Ounis, I.: University of glasgow at trec 2004: Experiments in web, robust and terabyte tracks with terrier. In: Proc. of the 13th Text Retrieval Conference, TREC (2004)
35. Porter, M.F.: An algorithm for suffix stripping. Program 3(14), 130–137 (1980)

36. Salton, G., Allan, J., Buckley, C.: Approaches to passage retrieval in full text information systems. In: Proc. of the 16th Annual International ACM SIGIR Conference on Research and Development in Information Retrieval, SIGIR (1993)
37. Santos, R., He, B., Macdonald, C., Ounis, I.: Integrating proximity to subjective sentences for blog opinion retrieval. In: Proc. of the 31st European Conference on IR Research, ECIR (2009)
38. Santos, R.L.T., He, B., Macdonald, C., Ounis, I.: Integrating proximity to subjective sentences for blog opinion retrieval. In: ECIR, pp. 325–336 (2009)
39. Skomorowski, J., Vechtomova, O.: Ad hoc retrieval of documents with topical opinion. In: Proc. of the 29th European Conference on IR Research, ECIR (2007)
40. Tan, S., Wang, Y., Cheng, X.: Combining learn-based and lexicon-based techniques for sentiment detection without using labeled examples. In: Proc. of the 31st Annual International ACM SIGIR Conference on Research and Development in Information Retrieval, SIGIR (2008)
41. Turney, P.D.: Thumbs up or thumbs down?: semantic orientation applied to unsupervised classification of reviews. In: Proc. of the 40th Annual Meeting on Association for Computational Linguistics (2001)
42. Zhang, Q., Wang, B., Wu, L., Huang, X.: Fdu at trec 2007: opinion retrieval of blog track. In: Proc. of the 16th Text Retrieval Conference, TREC (2007)
43. Zipf, G.K.: Human Behavior and the Principle of Least-Effort. Addison-Wesley, Reading (1949)

A Suite of Semantic Web Tools Supporting Development of Multilingual Ontologies

Maria Teresa Pazienza, Armando Stellato, and Andrea Turbati

Abstract. The multilingual aspects which characterize the (Semantic) Web and the constant demand for more understandable and easy-to-share forms of knowledge representation, push for a more "linguistically aware" approach to ontology development and foresees an environment where formal semantics could coexist with natural language, contributing to improve "shareability" of the content they describe. As a consequence ontologies should be enriched to both cover formally expressed conceptual knowledge as well as to expose content in a linguistically motivated fashion. In this paper we present a suite of tools, libraries and ontologies, ranging from ontology development to language resources access and management, supporting the development of multilingual ontologies. The contribution of this work, going beyond mere tool presentation, is two-fold: the presented tools implicitly embody a new way (methodology?) of rethinking the development of ontologies in terms of making their content easy reusable and comprehensible; moreover, they represent living proofs of software engineering principles associated to software reuse, documentation, modularity, interaction analysis, applied to the domain of Knowledge Management Software.

1 Introduction

Semantic Web ontologies represent the shared vocabularies through which machines can read and access content from the Web, or even communicate between them, to exchange information or cooperate for achieving some goal. This definition implicitly assumes that in an heterogeneous scenario like the whole WWW, the same concepts will be represented by the same ontologies and that, therefore, ontological models of data will be consistent; conversely, sensible effort will be put in trying to match these "not-so-shared" vocabularies. If that general assumption may hold true for reduced-size, very specific and data-oriented

Maria Teresa Pazienza · Armando Stellato · Andrea Turbati
ART Group, Dept. of Computer Science, Systems and Production
University of Rome, Tor Vergata
Via del Politecnico 1, 00133 Rome, Italy
e-mail: {pazienza,stellato,turbati}@info.uniroma2.it

G. Armano et al. (Eds.): Intelligent Information Access, SCI 301, pp. 109–135.
springerlink.com

ontologies (e.g. the WGS84 Geo Positioning RDF vocabulary[1], which contains only a few properties for describing latitude, longitude and point-in-space concepts), for larger domain descriptions, requiring different levels of abstraction and different perspectives depending on local needs, we expect to see several, different ontologies arise from independent organizations, often addressing overlapping domains.

Two issues then urge to be solved: first, facilitating people and automated systems in performing alignments between ontologies where they represent the same concepts and, secondly, make their vocabularies more explicit to humans, so that they can be re-used consistently in different scenarios and by different actors; in this sense, logical consistency may only help in restricting the range of possible interpretations which may be assigned to logical symbols, while common-sense human reasoning using these vocabularies may beneficiate a lot by the presence of clear and exhaustive documentation. Extensive use of Natural Language contents, providing free descriptions, synonymical expressions and translations in different idioms of the intended meaning of a vocabulary, appears thus as the most intuitive kind of documentation for data structures such as ontologies, dealing with representation of domains. Several efforts have been undertaken to cover different aspects of this problem, motivating the adoption of linguistic resources for enriching ontology vocabularies with natural language contents [25,32,34,31,15], showing useful applications exploiting these combined resources [2,30,6], providing standards for representing this enrichment/integration, like in SKOS[2] (Simple Knowledge Organization Systems) and in [4], and promoting the development of techniques for automating this task [26].

Objective of our research work, which moves in between the Ontology Engineering and Natural Language Processing areas, is to strongly integrate conceptual and linguistic knowledge to reduce the everlasting gap which exists between these two forms of knowledge representation, breaking down the barrier between what is known as the "world model" of intelligent systems, and what is the "world outside there", characterized by real documents written in natural language.

In this context, a suite of tools, libraries and ontologies dedicated to the development of multilingual ontologies will be presented. First, the Linguistic Watermark Ontology Suite and Java Library: a suite of ontologies for describing both linguistic resources and software interfaces for accessing their content, other than representing (multi)lingual information inside ontologies, and a java extensible library providing interfaces (and a few implementations) for covering all of the above tasks. Then, the OntoLing Framework will be showed: a portable extension for ontology development tools supporting manual and semi-automatic annotation of ontological data with information from different, heterogeneous linguistic resources. Lastly, we describe Semantic Turkey, a Web Browser extension for Knowledge Management and Acquisition of Semantic Web data, and introduce for the first time OntoLing-ST, an implementation of the recent OntoLing 4.0 which, thanks to its high portability across different platforms and ontology

[1] http://www.w3.org/2003/01/geo/wgs84_pos

[2] http://www.w3.org/TR/swbp-skos-core-guide/

standards, has been easily integrated in the Semantic Turkey environment. Before introducing the above tools, section 2 will discuss state-of-the-art on language representation and linguistic resource modeling, while section 3 exposes our desiderata in reconsidering the process of ontology development, and details requirements for building applications for multilingual ontology development.

2 State of the Art and Standards for Linguistic Resources and Language Representation

"The term linguistic resources refers to (usually large) sets of language data and descriptions in machine readable form, to be used in building, improving, or evaluating natural language (NL) and speech algorithms or systems" [8]. Examples of linguistic resources are written and spoken corpora, lexical databases, grammars, treebanks and field notes. In particular, this definition includes lexical databases, bilingual dictionaries and terminologies (which can all be indicated as lexical resources), which may reveal to be necessary in the context of a more linguistic-aware approach to KR. In past years, several lexical resources were developed and made accessible (a few for free), and a wide range of resources is now available, ranging from simple word lists to complex MRDs and thesauruses. These resources largely differentiate between the explicit linguistic information they expose, which may vary in format, content granularity and motivation (linguistic theories, task or system-oriented scope etc…).

Multiple efforts have been spent in the past towards the achievement of consensus among different theoretical perspectives and systems design approaches. The Text Encoding Initiative (www.tei-c.org) and the LRE-EAGLES (Expert Advisory Group on Linguistic Engineering Standards) project [5] are just a few, bearing the objective of making possible the reuse of existing (partial) linguistic resources, promoting the development of new linguistic resources for those languages and domains where they are still not available, and creating a cooperative infrastructure to collect, maintain, and disseminate linguistic resources on behalf of the research and development community.

A more recent effort is given by the Lexical Markup Framework [11] – which is now pursuing ISO standardization – a UML-based model for the description of Lexical Resources. However, at the present time, a definitive standard is not available. Often, even a local agreement on the model adopted to describe a given (a series of) resource does not prevent from an incorrect formulation of its content. This is due to the fact that many resources have been initially conceived for humans and not for machines. As an example, in existing available dictionaries, the definitions of words and synonyms are not always managed the same way: in some cases synonyms are clustered upon the senses which are related to the particular term being examined (among others, Babylon, www.babylon.com, and Dict, www.dict.org/bin/Dict dictionaries, where the senses are separated by a ";" symbol), other simply report flat lists of terms without even identifying their different meanings (as in Freelang dictionaries: www.freelang.com). In several dictionaries, synonyms are mixed with extended definitions (glosses) in an unpredictable way and it is not possible to automatically distinguish them. Terms reported as synonyms may sometimes not be truly synonyms of the selected term, but may represent more specific or general concepts (this is the case of the Microsoft Word synonymn prompter). Of course, the ones mentioned above represent mere dictionaries not adhering to any particular linguistic model, though they may represent valuable resources on their own.

A much stronger model is offered by WordNet [21,10], which, being a structured lexical database, presents a neat distinction between words, senses and glosses, and is characterized by diverse semantic relations like hypernymy/hyponymy, antonymy etc… Though not being originally realized for computational uses, and being built upon a model for the mental lexicon, WordNet has become a valuable resource in the human language technology and artificial intelligence. Due to its vast coverage of English words, WordNet provides general lexico-semantic information on which open-domain text processing is based. Furthermore, the development of WordNets in several other languages [37,33,35] extends this capability to trans-lingual applications, enabling text mining across languages.

3 Linguistic Enrichment of Ontologies: Motivation and Desiderata

Ontology Development is a task requiring considerable human involvement and effort, at a large extent with the objective of providing a shareable perspective over domain related knowledge. What "shareable" means, depends on the nature of the task(s) the ontology is thought for. The scenario offered by the Semantic Web is in fact characterized by distributed services which must both realize and rely on a proper connection of machine-accessible formal semantics and more traditional Web content.

For this connection to be true, a complete Ontology Development process should consider the formal aspects of conceptual knowledge representation, as well as guarantee that the same knowledge be recognizable amongst its multiple expressions which are available on real data: that is language.

To achieve such an objective, we should reconsider the process of Ontology Development to include the enrichment of semantic content with proper lexical expressions in natural language. Ontology Development tools should reflect this need, supporting users with dedicated interfaces for browsing linguistic resources: these are to be integrated with classic views over knowledge data such as class trees, slot and instance lists, offering a set of functionalities for linguistically enriching concepts and, possibly, for building new ontological knowledge starting from linguistic one.

By considering some of our past experiences [1,29,27] with knowledge based applications dealing with concepts and their lexicalizations, a few basic functionalities for browsing linguistic resources (from now on, LRs) emerged to be mandatory:

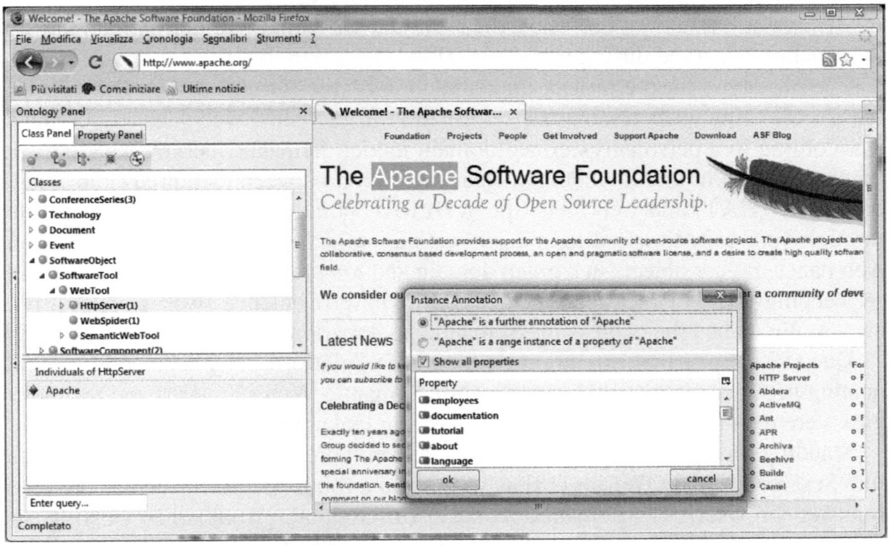

Fig. 1 Semantic Bookmarking with Semantic Turkey

- *Search term definitions (glosses)*
- *Ask for synonyms*
- *Separate different sense of the same term*
- *Explore genus and differentia*
- *Explore resource-specific semantic relations*

as well as some others for ontology editing:

- *Add synonyms (or translations, for bilingual resources) as additional labels for identifying concepts*
- *Add glosses to concepts description (documentation)*
- *Use notions from linguistic resources to create new concepts*

While ontologies have undergone a process of standardization which culminated, in 2004, with the promotion of OWL as the official ontology language for the semantic web, linguistic resources still maintain heterogeneous formats and follow different models, which make tricky the development of such an interface.

In the next sections we present our suite of tools for multilingual ontology development, starting by first through our ontology development and knowledge acquisition framework Semantic Turkey, and then presenting the suite of ontologies, software libraries and tools supporting multilingual enrichment of ontologies.

4 Semantic Turkey

Semantic Turkey [14] was born inside a national project – funded by the FILAS agency (Finanziaria Laziale di Sviluppo) under contract C5748-2005 – focused on innovative solutions for browsing the web and for collecting and organizing the information observed during navigation (Fig. 1).

The prototype for the project immediately took the form of a Web Browser extension allowing users to annotate information from visited web sites and organize it according to a personally defined domain model: Semantic Turkey paradigmatic innovation was in fact to "obtain a clear separation between (acquired) knowledge data (the WHAT) and web links (the WHERE)" pointing to it. That is, to be able, through very easy-to-use drag'n'drop gestures, to *select* textual information from web pages, *create* objects in a given domain and *annotate* their presence in the web by keeping track of the selected text and of its provenience (web page *url*, *title* etc…). We coined the expression "semantic bookmarking" for this kind of activity.

Due to its proverbial extendibility, the Firefox platform[3] had been chosen as the hosting browser for our application, while Semantic Web standards and technologies were the natural candidate for representing its knowledge model.

Standing on top of mature results from research on Semantic Web technologies, like Sesame [3] and OWLim [18] as well as on a robust platform such as the Firefox web browser, ST (Semantic Turkey) differentiates from other existing approaches which are more specifically tailored respectively towards knowledge management and editing [13], semantic mashup and browsing [9,16] and pure semantic annotation [7,17], by introducing a new dimension which is unique to the process of building new knowledge while exploring the web to acquire it.

By focusing on this aspect, which has been further investigated in the two years of finalization leading to the current release, we went beyond the original concept of Semantic Bookmarking and tried to amplify the potential of a new Knowledge Management and Acquisition System: we thus aimed at reducing the impedance mismatch between domain experts and knowledge investigators on the one side, and knowledge engineers on the other, providing them with a unifying platform for acquiring, building up, reorganizing and refining knowledge.

4.1 Semantic Turkey Architecture

The architecture (Fig. 2) of Semantic Turkey follows a three layered design, with the presentation layer embodying the true Firefox extension and the other two layers built around java technologies for administering the business logic and data access.

Everything relating user interaction is directly managed by the Firefox extension, thanks to a solution directly integrated in the browser. This approach has two main advantages: total reuse of the functionalities of a well assessed, stable and complete software for web browsing, and a non invasive offer for the user, who can still use the web browser he has been acquainted with.

The second layer, the service layer, is realized through a collection of Java Web Services, published through the Web Server "Jetty"[4]. Jetty is implemented entirely in Java, and the architecture foresees its use as an embedded component. This

[3] `http://www.mozilla.com/en-US/firefox/`

[4] `http://jetty.mortbay.org/jetty/`

means that the Web Server and the Web Application run in the same process, without interconnection overheads and other sort of complications.

The following sections describe more in detail the three layers which constitute the architecture of Semantic Turkey

Presentation Layer. The User Interface has been created through a combined use of the XML User Interface Language XUL[5], XBL[6] and Javascript language.

The UI physically appears as a set of Firefox sidebar, representing ontological information. User requests are handled through the Ajax [12] paradigm: the data – in XML format – is thus mainly exchanged between the two layers in an asynchronous way, to preserve good performance and to not penalize the activity of the browser.

Javascript XPCOM[7] components have been developed and the Simile Java Firefox Extension[8] has been adopted for linking the chrome part and the Java part to start the Jetty embedded java server.

Middle Layer. This layer offers services which may be invoked through http requests submitted according to the Ajax paradigm, thus enabling communication between the client (Firefox extension) and the server. The server receives the requests coming from the client by GET or POST http calls, carries out the operations associated to these calls, and in case replies with an XML response. If a call implies the return of a XHTML page, an XSLT transformation is being performed, in order to decouple the data model with its manifestation in the presentation layer.

The majority of invocations to the server are being completed in an asynchronous way, so that, independently from the workload that is subjected the server, the browser can continue to respond to the user. This is a crucial issue for the usability of the application: expensive computations blocking normal behavior of the browser would otherwise not be tolerated by the user.

Besides supporting the communication with the client, the middle layer provides the functionalities for definition, management and treatment of the data. Several objects are described through an ontological model (see next section), to represent both pure conceptual knowledge as well as application required information.

Data layer. It is mainly constituted by the component for managing the ontology. This has recently been rewritten as a series of dedicated API for accessing ontological data: these offer both RDF triple-level access methods as well as more object oriented facilities, which have been appreciated in RDF libraries like Jena [20]. Semantic Turkey API constitute an interface which can be implemented by building wrappers for existing ontology libraries, so that we could easily select those which best fit the needs of a given situation (like working with small or large repositories, on a local or collaborative environment etc…) without having to modify the whole application. The first implementation of these API has been

---

[5] `http://www.mozilla.org/projects/xul/`

[6] `http://www.mozilla.org/projects/xbl/xbl.html`

[7] `http://www.mozilla.org/projects/xpcom/`

[8] `http://simile.mit.edu/java-firefox-extension/`

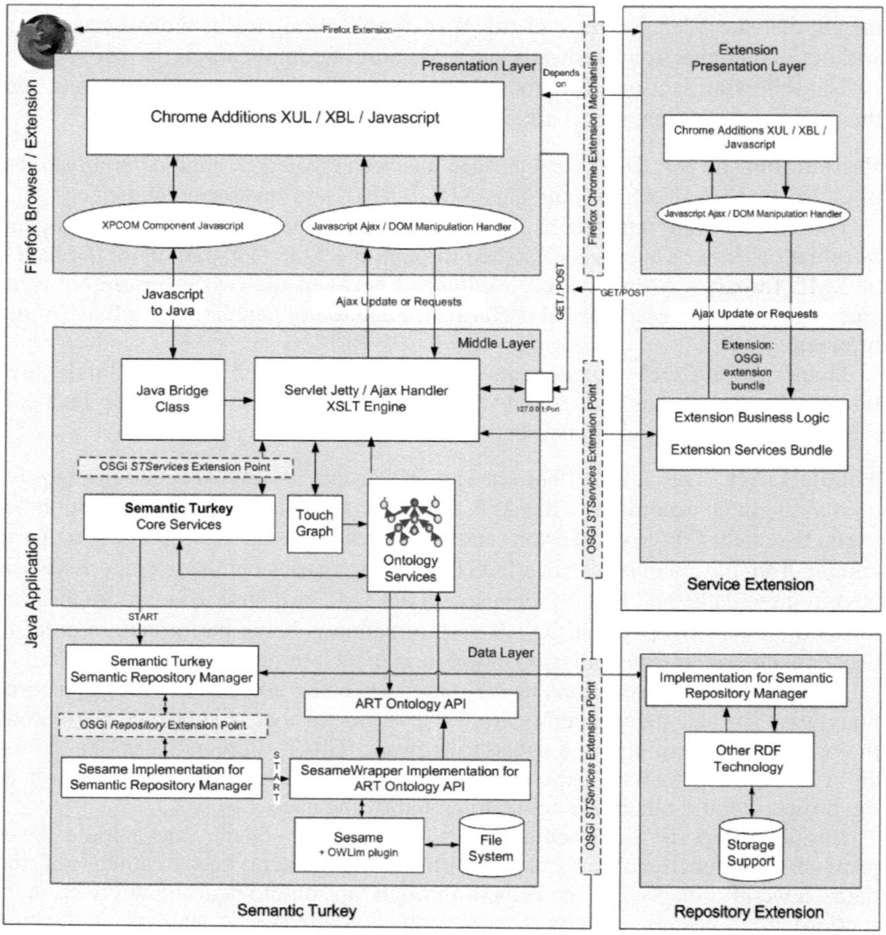

Fig. 2 Architecture of Semantic Turkey and of its extensions

developed as a wrapper for Sesame [3] and the OWLIM plugin [18], which has been added for reasoning over OWL [38] data.

Semantic Turkey also features an extension mechanism supporting both technologies belonging to the Front End and the Business and Data Layers.

The whole extension mechanism is obtained by a proper combination of the Mozilla extension framework (which is used to extend the user interface, drive user interaction and add/modify browser functionalities of ST) and the OSGi java extension framework [23] (providing extensions capabilities for the service and data layers of the architecture). OSGi compliance is obtained through the OSGi implementation developed inside the Apache Software Foundation, called Felix (felix.apache.org/).

Two main extension points have been introduced: a Service extension and a Repository Extension. The first one allows for the development of arbitrary

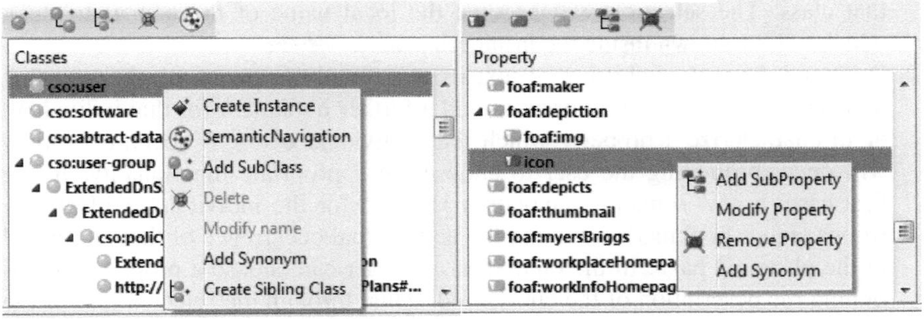

Fig. 3 Class and Property panels in Semantic Turkey

services which can be added dynamically to the system. Extensions of this type typically need to realize both a client extension through Mozilla technology, by adding new functionalities (and hooks for them in the user interface) to the system, as well the corresponding Service which is added dynamically through OSGi.

The second kind of extension provides openness to different triples store technologies; Semantic Turkey is in fact no more strictly based on the Sesame + OWLim libraries for RDF management, but features proprietary APIs for querying the managed ontologies. These API are defined through a set of interfaces, which can be implemented to adopt different triple stores. This can be of particular interest in specific scenarios where the target user has to connect to a specific triplestore, or where a service extension is being built by annexing an existing application, and in either case, these are based on a different triple store technologies.

Both kind of extensions are deployable as an xpi (cross-platform installers) packages which, once installed inside Firefox, are handled by Semantic Turkey extension discovery system, which extracts OSGi bundles and installs them in the main application. This assures easy installation for the user, which can install ST extensions as any other Firefox one, by dragging the xpi over Firefox and restarting the browser.

4.2 User Interaction

Semantic Turkey offers editing operations for populating the *personal ontology* with annotations from visited web sites, as well as search and navigation functionalities which facilitate the recovery of already acquired knowledge.

Main functionalities. The user may interact with the ontology panel to modify its personal ontology, through a series of operations, which we describe here, organized into categories.

Interaction with the browser. These mainly include drag&drop operations which allow to annotate information from the visited sites:

1. Drag and drop of a selection of a text from an html document displayed in the browser, on the icon that represents a class, in order to create an individual of

that class. The selection will become the local name of the new individual, which will be shown inside the instances panel.

2. Drag and drop of a selection of text from an html document, on the icon that represents an individual, in order to add a further bookmark for that individual, or to characterize a property which that individual owns. A specific window will open, prompting the user to choose the appropriate functionality. In the first case, a new semantic annotation is taken for the individual, with a new webpage as a bookmark for it and the new textual occurrence of that individual in the observed page. In the other case, the user can choose a property for enriching the description of the chosen individual through the selected text. If the selected property is an *owl:ObjectProperty*, the selection will become the name of a new individual created as an instance of the range class of the chosen property, or a further annotation for an existing individual. In both cases, the two individuals are bound through the selected property. In case of an *owl:Datatype* or *owl:AnnotationProperty*, a new value will be added.
3. Drag and drop of a selection of text from an html document, on the icon that represents an individual, in order to define a further lexicalization for that individual. The user can choose, from the same panel described before, if the selection characterizes a range of a property or a new *lexicalization*.

These functionalities have been conceived to speed up typical series of operations which characterize both the worlds of ontology development and semantic annotation. For example, the second one which has been described above performs, in case of an object property, the creation of a new instance, its annotation with the current web page and the assertion of a relationship between the new individual and the selected one, at the cost of just a drag&drop and a selection.

Direct Ontology Editing. These functionalities operate exclusively on the ontologies, as it should be important for the user to integrate the knowledge acquired through semantic bookmarking with information he could get through other media. All typical ontology editing operations (Fig. 3) are carried out through buttons and context menus associated to the nodes of the tree, in a way much similar to traditional ontology editing tools, like Protégé [13] or TopBraid Composer[9]. By offering complete interaction with the ontology via the XUL interface (instead of an HTML interface, like in Piggy-Bank), the user is not diverted from his current navigation (i.e. the main browser panel is still focused on the visited web page, which would otherwise be replaced by the HTML UI) and may, at the same time, maintain its attention over the observed web page. Extended support for *natural language descriptions* of ontology objects is also present in the system, allowing for explicit representations of the same objects through different synonimical expressions, or translation for different idioms, thus accounting for multilinguism. This is a further aspect to be distinguished from keeping track of the several ways in which ontology objects have been annotated over web pages, since this last is thought for addressing other phenomena, like acronyms, misspells and other idiosyncratic expressions.

---

[9] `http://simile.mit.edu/java-firefox-extension/`

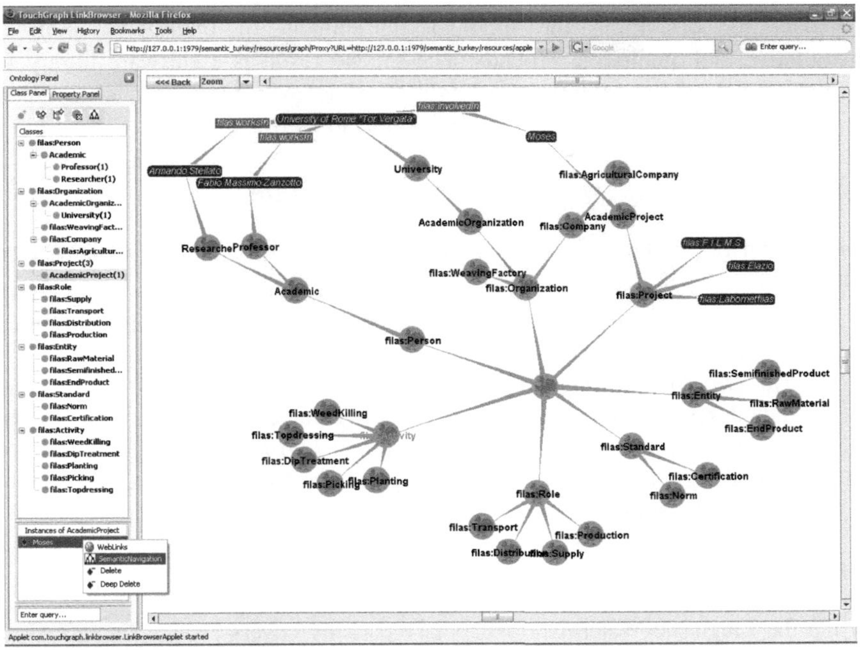

Fig. 4 Class and Property panels in Semantic Turkey

Semantic Browsing. As an additional feature, the user may graphically explore the ontology, thanks to the *SemanticNavigation* component: a customized version of the TouchGraph library[10] allowing for a graph-like exploration of ontology nodes. A Java applet will be loaded on a new tab of the browser, displaying the graph view of the ontology, allowing the user to navigate its content. The nodes of the graph will be displayed in different manners, according to the nature of the ontological entity: classes, properties or individuals. By dragging the mouse pointer on a node that represents an individual, it is possible to open a popup window, which contains the URLs of the pages where that instance has been annotated.

5 The Linguistic Watermark

The Linguistic Watermark [28] is an ontological and software framework for describing, referring and managing heterogeneous linguistic resources and for using their content to enrich and document ontological objects. It articulates into two results: first, a set of coordinated RDF vocabularies providing descriptors for representing linguistic resources (ranging from lexical to frame-based ones) and their software counterparts (data structures, access libraries etc...), as well as offering

[10] http://touchgraph.sourceforge.net/

metadata for describing the linguistic enrichment of ontologies, both on quantitative and qualitative grounds. The second result is a software library for evaluating the quality of automatic linguistic enrichment tools, through comparison of enriched ontologies compiled against the above vocabularies.

5.1 The Linguistic Watermark Ontology Suite

The Linguistic Watermark suite of RDF vocabularies is composed of three ontologies:

- The *Linguistic Watermark (LW)* vocabulary, describing linguistic resources through their purposes and structure organization
- *The Ontological Linguistic Watermark (OLW)* vocabulary: a set of metadata descriptors for characterizing the linguistic expressivity of ontologies
- *The LW Linguistic Interfaces vocabulary (LWLI)*, providing concepts for describing software libraries which grant access to specific (or ranges of) linguistic resources.

5.1.1 The Linguistic Watermark (LW) Vocabulary

While the Linguistic Watermark vocabulary partially covers general linguistic concepts like term, word, lexical/semantic relation, frame, agent etc... its main objective is to provide descriptors or characterizing the purpose and structure of linguistic resources: whether they represent translation vocabularies, synonyms collections, lexicons, frame based resources or terminologies, if they are organized around some kind of semantic structure or merely <entry, description> pairs etc..

Though originally conceived to cover any kind of Linguistic Resource, the first version of the Linguistic Watermark (Fig. 5) was limited to represent only lexical resources: by proper combination of its LW ontological descriptors, one could be able to represent very different linguistic resources, from simple synonym dictionaries, to complex resources such as WordNet [21]. This provided a shared and homogeneous vocabulary upon which multilingual (and multi-resource) applications could be defined.

In this work we have extended le LW vocabulary into two main directions:

- *RDF Porting*: now the LW model can be expressed as an RDF vocabulary
- *Instantiation*: now the vocabulary is not only used to describe linguistic resources, but even to predicate over their content (see section 4.2.2 for details).

Frames description: covering frame/class based linguistic resources, such as FrameNet and VerbNet (see [22] for further details).

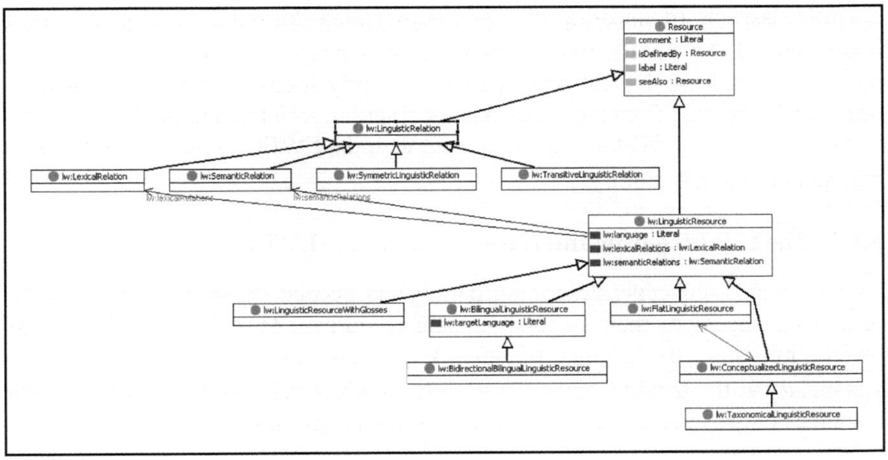

Fig. 5 An excerpt (focused on main descriptors for Linguistic Resources) from the Linguistic Watermark vocabulary

5.1.2 The Ontological Linguistic Watermark (OLW)

The characterization given by the OLW is expressed in terms of the linguistic content of the described ontology and with respect to the resources which have been adopted for enriching its concepts. As stated in [30], where its adoption has been considered in a scenario involving Semantic Coordination of FIPA agents, its metadata assume great significance in all the contexts where ontologies sharing a common domain, but no explicit semantic bridging between their respective vocabularies, need to be automatically aligned or merged. Resource-based algorithms for ontology alignment and semantic coordination agents can in fact inspect the OLW data of the ontologies to be compared and configure at best the resources and facilities to be used for matching their content. This is an aspect which has often been underestimated in literature: setting up the resources to be adopted in a realistic scenario, while being not a trivial task, influences dramatically the outcome and performances of any mediation activity.

The LWLI takes its roots from the first version of the Linguistic Watermark software library[11] – developed by the University of Rome, Tor Vergata – a component providing uniform access to different and heterogeneous linguistic resources, which has been used in several resource-based tools, such as the OntoLing Protégé plug-in [25] The LW presented in that work, was just a class diagram offering several interfaces and abstract classes whose combination could be used to describe the main aspects of a linguistic resource: implementing the proper subset of those (software) interfaces would result in the definition of a linguistic wrapper for accessing a particular linguistic resource. The LW library thus offered a combination of descriptive (with regard to the resources to be wrapped) and

[11] `http://art.uniroma2.it/software/LinguisticWatermark/`

operative aspects (delineating the operations which the required wrapper had to implement). Later on, the exigencies which brought to developing the OLW, required a formal ontological representation, merely focused on resource description, to be extracted from the original class diagram, which led to the LW.

Now, it was time to close the circle, and with the LWLI we recovered the original intent of the LW library.

5.1.3 The LW Linguistic Interfaces Vocabulary (LWLI)

LWLI contains concepts describing parameters needed by software libraries for setting up access to their target linguistic resources. This third ontology completely migrates the original framework to RDF, thus providing a complete vocabulary at the hand of Semantic Web tools which rely on the use of linguistic resources or are even expressly dedicated to the integration of ontologies with linguistic resources.

The LWLI includes concepts like:

- *LinguisticInterface*: for describing a specific implementation of a wrapper for a linguistic resource
- *LinguisticInterfaceConfiguration*: representing instances of basic runtime configurations for a given *LinguisticInterface*.
- *LinguisticInterfaceInstanceConfiguration*: each instance of this class provides data for completing a single runtime configuration for accessing a specific linguistic resource, basing on partial configuration from a given *LinguisticInterfaceConfiguration*.

and properties for specifying these configuration settings, among which, we list the following ones:

- *configuredInterface*: this property tells which *LinguisticInterface* is being configured through the described configuration
- *interfaceableResource*: tells which linguistic resources are made accessible through the described *Linguistic Interface*
- *ConfigurationProperty*: a property defining configuration parameters for accessing a linguistic resource through a dedicated linguistic interface. This property is never instantiated, though it has a few relevant subproperties for telling whether a given configuration parameter points to the file system, if a property is relevant for configuring a linguistic interface (InterfaceProperty) as a whole, or just for accessing specific resources (InstanceProperty) etc..

As for the LW, even this vocabulary provides an upper ontology which, though extensible in principle to match the specification of each represented software library, already contains all the required descriptors for automatically driving different linguistic resources under a shared knowledge model.

5.2 The Linguistic Watermark Library

Following the recent improvements on the LW suite, we are releasing a new version of the Linguistic Watermark library (LW 3.0), which offers java API for accessing linguistic resources through dedicated Linguistic Interfaces, both entities being defined according to the LW and LWLI vocabularies. In particular, a mapping between the above ontologies and newly added java interfaces allows implemented java wrappers for linguistic resources to declare themselves as new instances of the LinguisticInterface class and accept strongly typed configuration parameters, thus enabling data consistency checks and providing hooks for automatic generation of configuration user interfaces for hosting applications.

To implement this mechanism we adopted and OSGi compliant java extension framework: Apache Felix (felix.apache.org/). Each OSGi bundle (the OSGi name given to the extension packages) contains a class that extends the abstract class LIFactory (see class diagram in Fig. 6), which is in charge of generating objects implementing the LinguisticInterface interface. Each class that implements the LinguisticInterface interface has some of its fields representing specific InterfaceProperty and InstanceProperty properties (they are automatically identified through *java annotations*). InterfaceProperties share their value among all the instances, so they are declared as static fields, while InstanceProperties have values specific to each object (identifying a specific linguistic resource present in the host). LIFactories release new instances of LinguisticInterface by getting their needed configuration (i.e. InterfaceProperties and InstanceProperties values), which is stored in a LinguisticResource object, from a loaded LW LingModel. We implemented two serializations (and related loaders/writers) of the LingModel: one compact xml represention (handled my LingModelXMLIO) and an RDF representation which follows the LW RDF Vocabulary (LingModelRDFIO).

While there should be exactly one LinguisticInterface which is responsible for providing access to a specific loaded resource, proper handling of the LIFactory/LinguisticInterface pair can hide implementation issues related to wrapping and reusing existing foreign libraries with different architectures into this framework.

As an example, one existing library for a particular kind of resource – let us call it *LRESLIB* – could adopt one singleton object (*ResManager*) for managing different linguistic resources of the same type (different versions or for different languages). In this case, the *LRESLIB* library can be easily wrapped in the LW framework by initializing, storing and hiding *ResManager* inside its built LIFactory implementation, while the associated LinguisticInterface implementation will represent simple objects retaining reference to their LIFactory and invoking *ResManager* methods (with parameters customized for their specific resource) through delegation.

This approach guarantees reuse of existing libraries and tools for accessing linguistic resources while porting their provided content inside an extensible framework with well defined model, vocabulary and operations.

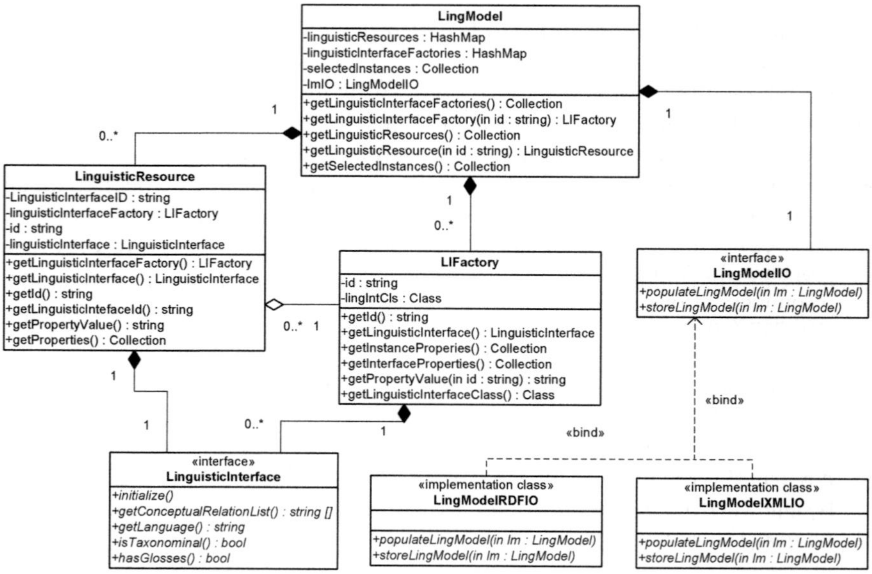

Fig. 6 Class diagram of the main components of LW model

5.2.1 The OLW Library and OLW Vocabulary Improvements

With the specific aim of obtaining a stable range of instruments for enriching ontologies with lexical content, and of formalizing the model and associated format for representing this information, we have developed a dedicated component which, together with the LW library, can be embedded in ontology based tools and applications needing to incorporate linguistic content.

The OLW Integration Model

In modeling our framework for the integration of ontological and linguistic content, we have taken into consideration the following requisites, which should allow for:

- Reporting quantitative and qualitative information on the overall process of enriching an ontology with content from a linguistic resource (this was the primary objective of the OLW metadata ontology)
- Keeping track (at least maintain the possibility to do that) of the source used for enriching the content
- Being able to properly map different kind of linguistic entities (words, linguistic/semantic relations etc…) with (structures of) ontological objects
- Giving the user the possibility of adopting resources' specific objects (e.g. FrameNet frames or WordNet synsets) for enriching an ontology

- Embedding existing models for integration of ontologies and linguistic entities, still respecting the above priorities
- Assessing reliable links between ontological and linguistic objects as well as taking into account for probabilistic matches produced by automatic enrichment tools (which could also be used for evaluation purposes)

The first requisite has been satisfied by defining a set of meta-descriptors – represented through object properties with domain set to owl: Ontology – for providing an overview of the "linguistic expressiveness" of ontologies. These properties may prove to be helpful for services/agents which, having to map/merge/align/mediate different ontologies, may be willing to invoke the proper linguistic resources for supporting this task. These mediators can thus beneficiate of the overall statistical information provided by the OWL metadata, without inspecting the entire ontologies' content. This part of the OLW has already been described in details in [30].

The second, third and fourth requisites have been accomplished by extending the LW; in its first incarnation, which served solely as a conceptual driver for the software library, the LW was able to express descriptions of linguistic resources, without predicating about their specific content. Now it has been extended to make possible the instantiation of objects from the described resources. The example in Fig. 7 shows fragments originating from three different ontologies: the first fragment is a description of WordNet synset 100001740 originating from the WordNet-RDF vocabulary developed by the WordNet task force of the W3C (http://www.w3.org/TR/wordnet-rdf/); the second one is the binding of concept wn20schema:Synset to the lw:SemanticIndex, through a rdfs:subClassOf relationship. Finally, a certain Noun concept coming from a fictitious ontology is enriched with the meaning expressed by the above synset, through the owl:semanticDescriptor property. With this extensible pattern, the LW+OLW offer reusable vocabularies for describing linguistic resources which drive the behavior of software applications serving the same task, while specific extensions (both in terms of ontologies and software components) can be added to describe specific lexical and semantic objects from new resources, without requiring modifications to the core vocabulary nor to the original application.

Compatibility with existing (proposed) models As previously mentioned, several formats exists or have been proposed for integrating ontological content with linguistic information.

While we did not intend to propose a new one, we tried to obtain cross-compatibility with available standards and proposed models, by gearing our software library with a OntoLinguisticModel interface, consisting of a series of enrichment/retrieval operations defined upon abstract "slots" for representing linguistic information. These slots can be then implemented according to a specific onto-linguistic representation model, by specifying the properties and concepts used to map/integrate linguistic information with ontological one.

```
<wn20schema:NounSynset rdf:about="wn20instances:synset-entity-noun-1" rdfs:label="entity">
        <wn20schema:synsetId>100001740</wn20schema:synsetId>
</wn20schema:NounSynset>

<rdf:Description rdf:about="wn20schema:Synset">
    <rdfs:subClassOf rdf:resource="lw:SemanticIndex"/>
</rdf:Description>

<someOntology:Noun>
        <olw:semanticDescriptor rdf:resource="wn20instances:synset-entity-noun-1">
</someOntology:Noun>
```

Fig. 7 An example of resource wrapping: binding WordNet-RDF synsets to a class concept

Obviously, it is impossible to foresee in advance all the characteristics of each model/interface-implementation which could be integrated in the future, thus we provided a specific *project/decode* feature for projecting the linguistic information extracted from linguistic resources according to the LW ontology, towards the (possibly more fine-grained) adopted ontolinguistic model. For evaluative (see next section) and comparative purpose in general, we demand to each specific implementation the specifications of equivalence between the locally defined linguistic objects.

Implementations of OntoLinguisticModel have been developed for the traditionally adopted RDFS annotation properties (rdfs:label and rdfs:comment), for the base SKOS vocabulary (by extending the above with skos:prefLabel and skos:altLabel), for SKOS +SKOS-Mapping[12] vocabularies (thus including skos:broader/skos:narrower and skos:related, to map ontology concepts with instances of lw:SemanticIndex from the LW ontology) and, finally, for the LingInfo model, by wrapping the linginfo:linginfo property and linginfo:LingInfo class. The above integration model satisfied our fifth requirement, while the resolution of the sixth one is part of the discussion presented in the next section.

5.3 The Evaluation Framework

The newly developed OLW Library provides a framework for evaluating the quality of algorithms for Linguistic Enrichment of ontologies with respect to previously defined reference standards, by using standard *precision&recall* metrics [36].

The OLW library can accept pairs of linguistic enrichment documents (that is: ontologies with integrated linguistic content), where one is the Oracle and the

[12] http://www.w3.org/2004/02/skos/mapping/spec/

other one is the result to be tested, providing that the following extensions are included in the library and properly configured:

- *Enrichment Model* and related software extension
- *Resource(s) description* (and their wrapper implementation) used for enrichment
- *Match Specification and Evaluation (MSE)* extension, if different enrichment entries differ from simple links between ontological and linguistic objects

With the ones above, the library is able to seek the enrichment properties (at least, those which need to be considered) in the ontology documents (first extension) and to properly identify the elements used for the enrichment (second extension). The third one is an extension needed for those cases where an algorithm produces any kind of probabilistic/quantitative result, so that the enrichment links in the tested document cannot be evaluated just in terms of correct/wrong matches versus those in the Oracle. Inter-annotator agreement can as well be measured against two enrichment documents compiled by human annotators, with no further requirement apart from above.

5.3.1 OntoLing

OntoLing [24] is, in its last incarnation (OntoLing 4.0), a generic architecture for extending Ontology Development tools with functionalities for enriching ontological knowledge with linguistic content. The architecture of OntoLing will be implementable through realization and composition of different components:

By first a core component exposing the following characteristics:

- can be *interfaced* with the Linguistic Watermark software library to access linguistic resources, and with different enrichment algorithms and models (see Linguistic Watermark description in previous section) for enriching the content of ontologies with information gathered from loaded resources.
- *knowledge* of the main functionalities and user interfaces characteristics exposed by common ontology development tools and of the extensions which should be brought by the OntoLing framework
- *high portability*: the core component has a module called `UIReasoner` (*User Interface Reasoner*) which is able to describe – according to an abstract representation formalism – the way the UI should appear to the user (which depends on the characteristics of the loaded linguistic resource) as well as describe actions and events which happen inside it. This way, a concrete implementation of this component could be easily ported and reused across different development environments. Moreover, if the abstraction layer is sufficiently expressive, changes to the core component should not require (heavy) modifications on each of its multiple implementations available for current ontology development tools.

Second, trivially: the Linguistic Watermark library

Third: a set of linguistic resources (and wrappers for them, compatible with Linguistic Watermark API)

Fourth: an ontology development tool

Fifth (and last), an adapter between OntoLing core component and the ontology development tool, which directly wraps its API and provides concrete implementations for OntoLing User Interface extensions.

5.4 OntoLing Core Application

The core component of the architecture is responsible for interpreting the Watermark of linguistic resources and for exposing those functionalities which suit to their profile. Moreover, the behavior of the whole application is dependent on the nature of the loaded resource and is thus defined at run-time. Several methods for querying LRs and for exposing results have been encapsulated into objects inside a dedicated library of behaviors: when a given LR is loaded, the core module parses its Linguistic Watermark and assigns specific method-objects to each GUI event.

With such an approach, the user is provided with a uniform view over diverse and heterogeneous linguistic resources, as they are described in the Linguistic Watermark ontology, and easily learns how to interact with them (thus familiarizing with their peculiarities) by following a policy which is managed by the system.

For example, with a flat resource, a search on a given term will immediately result in a list of (potential) synonyms inside a dedicated box in the GUI; instead, with a conceptualized resource, a list of word senses will appear in a results table at first, then it will be browsed to access synonymical expressions related to the selected sense. Analogous adaptive approaches have been followed for many other aspects of the Linguistic Watermark (mono or bidirectional Bilingual Translators, presence of glosses, Taxonomical structures and so on…) sometimes exploding with combinatorial growth.

Future development of Ontoling will go in the direction of considering supervised techniques for automatic ontology enrichment; selecting and modeling the right strategies for the adopted LRs is another task the core module is in charge for.

5.5 OntoLing User Interface

Once activated, the plug-in displays two main panels, the Linguistic Browser on the left side, and the Ontology Panel on the right side (see Fig. 9).

The Linguistic Browser is responsible for letting the user explore the loaded linguistic resource. Fields and tables for searching the LR and for viewing the results, according to the modalities decided by the core component, are made available. The menu boxes on the left of the Linguistic Browser are filled at run time with the methods for exploring LR specific Lexical and Conceptual relations.

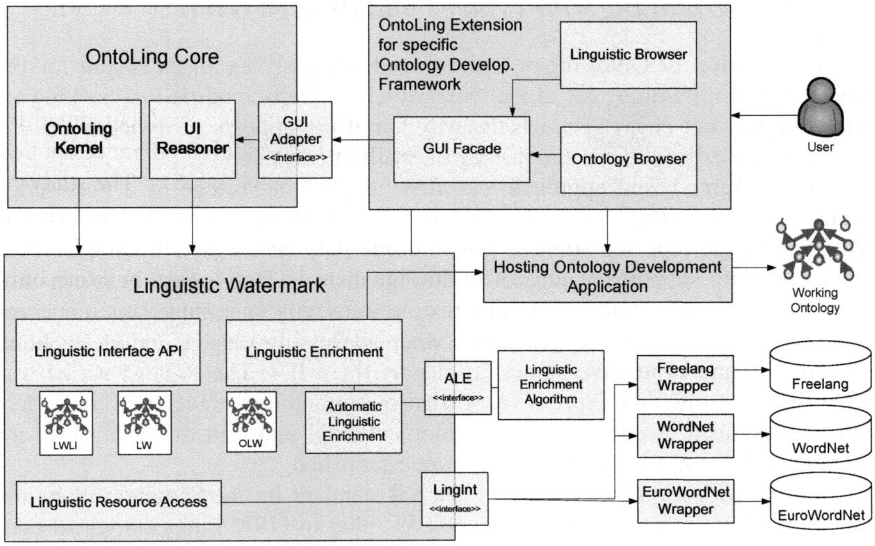

Fig. 8 OntoLing Framework Architecture

The Ontology Panel, on the right, offers a perspective over ontological data in the classic Protégé style. By right-clicking on a frame (class, slot or instance), the typical editing menu appears, with some further options provided by OntoLing to:

1. search the LR by using resources names as keys
2. change the name of the selected resource by using a term selected from the Linguistic Browser
3. add terms selected from the Linguistic Browser as additional labels for the selected resource
4. add glosses as a description (*rdfs:comment*) for the selected resource
5. add IDs of senses selected from the linguistic browser as additional labels for the resources
6. create a new resource with a term selected from the Linguistic Browser as resource name
7. only in class and property browser: if the LR is a `TaxonomicalLR`, explore hyponyms (up to a chosen level) of the concept selected on the Linguistic Browser and reproduce the tree on the resource browser, starting from the selected resource, if available

These functionalities allow not only for linguistic enrichment of ontologies, but can be helpful for Ontologists and Knowledge Engineers in creating new ontologies or in improving/modifying existing ones.

In OntoLing-Protégé, how terms and glosses are added to the description of ontologies concepts, depends on the ontology model which is being adopted and is explained in detail in the following section.

5.6 Using OntoLing with Protégé and Protégé OWL

The first version of OntoLing was developed expressly as an extension for the Protégé Ontology Editor. All of the work which radically modified its backing architecture has not changed much the way OntoLing appears to its users. In this section we describe choices and history of this first extension.

When a frame-based approach was first adopted in Protégé as a knowledge model for representing ontologies and knowledge bases, no explicit effort was dedicated to the representation of possible alternate labels (synonyms) for concepts neither to support the idea of multilingualism in Ontologies. Frame names were almost as equivalent as IDs, and people were only encouraged, as it is common practice in computer programming when addressing variable names, to adopt "meaningful and expressive names" to denote these IDs. The Protégé model was indeed quite strong and expressive, so that every ontology developer could deal with his linguistic needs at a meta-ontological level and find the right place for them, though no official agreement was yet established.

Later on, with the advent of OWL as a KR standard for the Semantic Web, and with the official release of the Protégé OWL plug-in [19], things started to converge towards a minimal agreement for the use of language inside ontologies. When we first started working on OntoLing, the OWL plug-in had just been released, and the majority of users continued to use Protégé in the usual way, so we had to find a solution that was quite easy (for the user) to make do with this lack in the standard Protégé model.

To this end, we defined the notion of terminological slot, as a slot which is elected by the user to contain different linguistic expressions for concepts. Any string-typed slot with cardinality set to multiple, can potentially be selected as a terminological slot, and, for easiness of use, OntoLing prompts the user only with this class of slots. This way, to use Ontoling with standard Protégé, a user only needs to define a proper metaclass and metaslot, containing the elected terminological slot; naturally, the same slot can be dedicated to instances at class level. Multilingual ontologies can also be supported by creating different slots and selecting each of them as terminological slots during separate sessions of Linguistic Enrichment, with diverse LRs dedicated to the different chosen languages. Concerning glosses, these can be added to the common "documentation" slot which is part of every frame by default.

Conversely, Linguistic Enrichment of OWL Ontologies follows a more predictable path, thanks to OWL's language dedicated Annotation Properties, such as *rdfs:label* and *owl:comment*. When Ontoling recognizes a loaded ontology as expressed in the OWL language, the terminological slot is set by default (though modifiable) to *rdfs:label*. In this case the *xml:lang* attribute of the label property is automatically filled with the language declared by the Linguistic Interface.

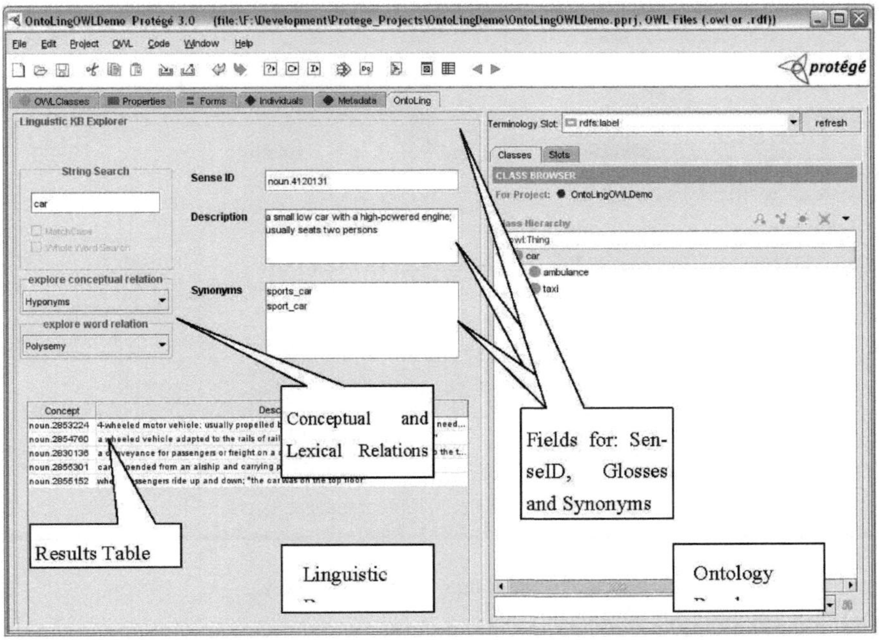

Fig. 9 OntoLing Screeshot (Protégé version)

5.7 OntoLing as an Extension to Semantic Turkey

While recent changes to the architecture of OntoLing have not produced (they were not meant do that) sensible impact on interaction with the user, they surely allowed for more flexible development of new functionalities as well as fast-to-produce porting over different applications.

Our experience in porting the new version of OntoLing on the Semantic Turkey architecture revealed that we were able to keep down realization costs by more than two thirds of the whole development effort, since we had to:

- realize its user interface
- realize a ST service extension which includes the OntoLing Core component
- serialize abstract UI actions produced by OntoLing Core component as XML messages sent from ST server
- develop handlers for UI actions sent by the server, realizing necessary handling of requested actions over the Firefox UI of OntoLing

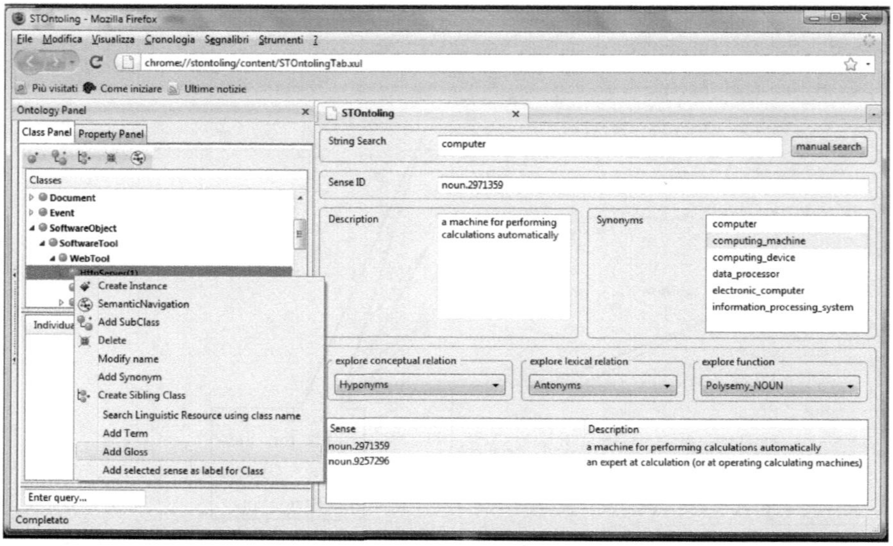

Fig. 10 Ontoling in Semantic Turkey

Of the above, only the first part required sensible effort, due to the completely different UI technology adopted by Firefox with respect to traditional Java Swing adopted by Protégé, thus preventing even minimal reuse of code. On the other hand, this aspect is a necessary step for any porting attempt, while we totally beneficiated of the complex UI management (depending on the ling. watermark of the loaded resource) which has been completely demanded to the included core component. Also, apart from the effort, this approach is not requiring deep knowledge of the framework nor of its inner logic, since most relevant and critical aspects are concentrated inside the core component and need not to be re-implemented: this lowers requirements in terms of development personnel and eases even more the porting process.

Though we focused in obtaining a portable and completely replicable multilingual extension for Ontology Development systems, we plan to obtain the best from the combination of OntoLing with the possibilities of our ontology development environment, deriving from its inherent connection with the Web and, as a consequence of that, with the many different information sources (Wikipedia, online dictionaries etc...) which can be explored in such an open environment.

6 Conclusion

In this paper we presented a collection of software libraries, tools and ontologies for supporting multilingual development of Semantic Web ontologies. The presented work is the result of different research efforts which we tried to converge towards a common goal, though this can be seen just as "end of the beginning" of this exploration.

We expect that our work, through its tangible proofs-of-concepts, may give a contribution or at least motivate the standardization of models, methodologies and tools for the effective integration of ontologies and linguistic resources: something which is much felt as a need for the future of Web 3.0 – which on the one side foresees a web of data made accessible by machines, and on the other one expects this data to be self-explanatory and human-comprehensible on a multicultural and multilingual ground – but which is until now demanded to specific efforts and arbitrary solutions.

References

[1] Atzeni, P., et al.: Ontology-based question answering in a federation of university sites: The MOSES case study. In: Meziane, F., Métais, E. (eds.) NLDB 2004. LNCS, vol. 3136, pp. 413–420. Springer, Heidelberg (2004)

[2] Basili, R., Vindigni, M., Zanzotto, F.M.: Integrating Ontological and Linguistic Knowledge for Conceptual Information Extraction. In: IEEE/WIC International Conference on Web Intelligence, Washington, DC, USA (2003)

[3] Broekstra, J., Kampman, A., van Harmelen, F.: Sesame: A Generic Architecture for Storing and Querying RDF and RDF Schema. In: The Semantic Web - ISWC 2002: First International Semantic Web Conference, Sardinia, Italy, June 9-12, pp. 54–68 (2002)

[4] Buitelaar, P., et al.: LingInfo: Design and Applications of a Model for the Integration of Linguistic Information in Ontologies. In: OntoLex 2006, Genoa, Italy (2006)

[5] Calzolari, N., McNaught, J., Zampolli, A.: EAGLES Final Report: EAGLES Editors Introduction, Pisa, Italy, EAG-EB-EI (1996)

[6] Cappelli, A., Giovannetti, E., Michelassi, P.: Ontological Knowledge and Language in Modelling Classical Architectonic Structures. In: Ontology and Lexical Resources - OntoLex 2004, Lisboa, Portugal (2004)

[7] Ciravegna, F., Dingli, A., Petrelli, D., Wilks, Y.: User-system cooperation in document annotation based on information extraction. In: Gómez-Pérez, A., Benjamins, V.R. (eds.) EKAW 2002. LNCS (LNAI), vol. 2473, p. 122. Springer, Heidelberg (2002)

[8] Cole, R.A., et al. (eds.): Survey of the State of the Art in Human Language Technology. Cambridge University Press, Cambridge (1997)

[9] Dzbor, M., Domingue, J., Motta, E.: Magpie: Towards a Semantic Web Browser. In: 2nd International Semantic Web Conference (ISWC 2003), Florida, USA (2003)

[10] Fellbaum, C.: WordNet: An Electronic Lexical Database. WordNet Pointers, MIT Press, Cambridge (1998)

[11] Francopoulo, G., et al.: Lexical Markup Framework (LMF). In: LREC 2006, Genoa, Italy (2006)

[12] Garrett, J.J.: Ajax: A New Approach to Web Applications (February 2005), http://www.adaptivepath.com/publications/essays/archives/000385.php

[13] Gennari, J., et al.: The evolution of Protégé-2000: An environment for knowledge-based systems development. International Journal of Human-Computer Studies 58(1), 89–123 (2003)

[14] Griesi, D., Pazienza, M.T., Stellato, A.: Semantic Turkey - a Semantic Bookmarking tool (System Description). In: Franconi, E., Kifer, M., May, W. (eds.) ESWC 2007. LNCS, vol. 4519, pp. 779–788. Springer, Heidelberg (2007)

[15] Huang, C.: Sinica BOW: Integrating bilingual WordNet and SUMO Ontology. In: Ontology and Lexical Resources - OntoLex 2004, Lisboa, Portugal (2004)

[16] Huynh, D., Mazzocchi, S., Karger, D.R.: Piggy Bank: Experience the Semantic Web Inside Your Web Browser. In: Fourth International Semantic Web Conference (ISWC 2005), Galway, Ireland, Galway, Ireland, November 2005, pp. 413–430 (2005)

[17] Kahan, J., Koivunen, M.-R.: Annotea: an open RDF infrastructure for shared Web annotations. In: WWW 2001: Proceedings of the 10th international conference on World Wide Web, Hong Kong, Hong Kong, pp. 623–632 (2001)

[18] Kiryakov, A., Ognyanov, D., Manov, D.: OWLIM – a Pragmatic Semantic Repository for OWL. In: Int. Workshop on Scalable Semantic Web Knowledge Base Systems (SSWS 2005), WISE 2005, New York City, USA, November 20 (2005)

[19] Knublauch, H., Fergerson, R.W., Noy, N.F., Musen, M.A.: The Protégé OWL Plugin: An Open Development Environment for Semantic Web Applications. In: Third International Semantic Web Conference - ISWC 2004, Hiroshima, Japan (2004)

[20] McBride, B.: Jena: Implementing the RDF Model and Syntax Specification. In: Semantic Web Workshop, WWW 2001 (2001)

[21] Miller, G.A., Beckwith, R., Fellbaum, C., Gross, D., Miller, K.: Introduction to WordNet: An On-line Lexical Database (1993)

[22] Oltramari, A., Stellato, A.: Enriching Ontologies with Linguistic Content: an Evaluation Framework. In: The role of ontolex resources in building the infrastructure of Web 3.0: vision and practice (OntoLex 2008), Marrakech, Morocco, May 31 (2008)

[23] OSGi RFC0112 (2005),
`http://www2.osgi.org/Download/File?url=/download/rfc-0112_BundleRepository.pdf`

[24] Pazienza, M.T., Stellato, A.: An open and scalable framework for enriching ontologies with natural language content. In: Ali, M., Dapoigny, R. (eds.) IEA/AIE 2006. LNCS (LNAI), vol. 4031, pp. 990–999. Springer, Heidelberg (2006)

[25] Pazienza, M.T., Stellato, A.: Exploiting Linguistic Resources for building linguistically motivated ontologies in the Semantic Web. In: Second Workshop on Interfacing Ontologies and Lexical Resources for Semantic Web Technologies (OntoLex2006), Genoa, Italy (2006)

[26] Pazienza, M.T., Stellato, A.: Linguistic Enrichment of Ontologies: a methodological framework. In: Second Workshop on Interfacing Ontologies and Lexical Resources for Semantic Web Technologies (OntoLex2006), Genoa, Italy (2006)

[27] Pazienza, M.T., Stellato, A., Enriksen, L., Paggio, P., Zanzotto, F.M.: Ontology Mapping to support ontology-based question answering. In: Second MEANING workshop, Trento, Italy (February 2005)

[28] Pazienza, M.T., Stellato, A., Turbati, A.: Linguistic Watermark 3.0: an RDF framework and a software library for bridging language and ontologies in the Semantic Web. In: Semantic Web Applications and Perspectives, 5th Italian Semantic Web Workshop (SWAP2008), FAO-UN, Rome, Italy, December 15-17 (2008)

[29] Pazienza, M.T., Stellato, A., Vindigni, M., Valarakos, A., Karkaletsis, V.: Ontology integration in a multilingual e-retail system. In: HCI International 2003, Crete, Greece (2003)

[30] Peter, H., Sack, H., Beckstein, C.: SMARTINDEXER – Amalgamating Ontologies and Lexical Resources for Document Indexing. In: Workshop on Interfacing Ontologies and Lexical Resources for Semantic Web Technologies (OntoLex2006), Genoa, Italy (2006)
[31] Philpot, A., Hovy, E., Pantel, P.: The Omega Ontology. In: Ontology and Lexical Resources (OntoLex2005), Jeju Island, South Korea (2005)
[32] Prevot, L., Borgo, S., Oltramari, A.: Interfacing Ontologies and Lexical Resources. In: OntoLex2005 - Ontologies and Lexical Resources, Jeju Island, South Korea (2005)
[33] Roventini, A., et al.: ItalWordNet: A Large Semantic Database for the Automatic Treatment of the Italian Language. In: First International WordNet Conference, Mysore, India (January 2002)
[34] Scheffczyk, J., Baker, C.F., Narayanan, S.: Ontology-based Reasoning about Lexical Resources. In: Workshop on Interfacing Ontologies and Lexical Resources for Semantic Web Technologies (OntoLex2006), Genoa, Italy (2006)
[35] Stamou, S., et al.: BALKANET: A Multilingual Semantic Network for the Balkan Languages. In: First International Wordnet Conference, Mysore, India, January 2002, pp. 12–14 (2002)
[36] Van Rijsbergen, C.J.: Information Retrieval. Butterworths, London (1975)
[37] Vossen, P.: EuroWordNet: A Multilingual Database with Lexical Semantic Networks. Kluwer Academic Publishers, Dordrecht (1998)
[38] W3C, `http://www.w3.org/TR/owl-features/`

Author Index

GPSR Compliance

The European Union's (EU) General Product Safety Regulation (GPSR) is a set of rules that requires consumer products to be safe and our obligations to ensure this.

If you have any concerns about our products, you can contact us on ProductSafety@springernature.com

In case Publisher is established outside the EU, the EU authorized representative is:

Springer Nature Customer Service Center GmbH
Europaplatz 3
69115 Heidelberg, Germany

Batch number: 10614853

Printed by Printforce, the Netherlands